Urban Sustainability

The Urban Sustainability Book Series is a valuable resource for sustainability and urban-related education and research. It offers an inter-disciplinary platform covering all four areas of practice, policy, education, research, and their nexus. The publications in this series are related to critical areas of sustainability, urban studies, planning, and urban geography.

This book series aims to put together cutting-edge research findings linked to the overarching field of urban sustainability. The scope and nature of the topic are broad and interdisciplinary and bring together various associated disciplines from sustainable development, environmental sciences, urbanism, etc. With many advanced research findings in the field, there is a need to put together various discussions and contributions on specific sustainability fields, covering a good range of topics on sustainable development, sustainable urbanism, and urban sustainability. Despite the broad range of issues, we note the importance of practical and policy-oriented directions, extending the literature and directions and pathways towards achieving urban sustainability.

The series will appeal to urbanists, geographers, planners, engineers, architects, governmental authorities, policymakers, researchers of all levels, and to all of those interested in a wide-ranging overview of urban sustainability and its associated fields. The series includes monographs and edited volumes, covering a range of topics under the urban sustainability topic, which can also be used for teaching materials.

Yu Bo · Shida Sun · Liling Wu · Kebin He

Pathways for Reaching Air Quality Standard and Carbon Peak in Selected Cities in China

中国工信出版集团 電子工業出版社 PUBLISHING HOUSE OF ELECTRONICS INDUSTRY http://www.phei.com.cn Springer

Yu Bo
Institute of Atmospheric Physics
Chinese Academy of Sciences
Beijing, China

Shida Sun
Department of Earth System Science
Tsinghua University
Beijing, China

Liling Wu
School of Environment
Tsinghua University
Beijing, China

Kebin He
School of Environment
Tsinghua University
Beijing, China

ISSN 2731-6483 ISSN 2731-6491 (electronic)
Urban Sustainability
ISBN 978-981-96-8314-7 ISBN 978-981-96-8315-4 (eBook)
https://doi.org/10.1007/978-981-96-8315-4

Jointly published with Publishing House of Electronics Industry
The print edition is not for sale in China (Mainland). Customers from China (Mainland) please order the print book from: Publishing House of Electronics Industry.

Translation from the Chinese Simplified language edition: "典型城市空气质量达标及碳排放达峰路径研究" by Yu Bo et al., © Publishing House of Electronics Industry 2023. Published by Publishing House of Electronics Industry. All Rights Reserved.

This Springer imprint is published by the registered company Springer Nature Singapore Pte Ltd.
The registered company address is: 152 Beach Road, #21-01/04 Gateway East, Singapore 189721, Singapore

Contents

Chapter 1
Background

Abstract This chapter provides a succinct overview of China's advancements in addressing atmospheric pollution and climate change, alongside an analysis of its challenges in meeting urban air quality standards and achieving carbon emission peaks. The implementation of initiatives such as the "Air Pollution Prevention and Control Action Plan" and the "Three-Year Action Plan for Winning the Blue Sky Defense Battle" has notably bolstered China's environmental air quality. However, fundamental restructuring to address structural pollution remains incomplete, as evidenced by the fact that in 2020, forty percent of cities nationwide still exceeded established air quality thresholds. Moreover, with the diminishing efficacy of end-of-pipe control measures and narrowing avenues for pollutant reduction, sustaining air quality improvements presents an increasingly daunting task, necessitating exploration of new avenues for emission mitigation. Regarding peak carbon emissions, the evolution of China's economy and urbanization has entrenched high-carbon energy and industrial frameworks as pivotal determinants. The interconnectedness of greenhouse gases and atmospheric pollutants in terms of their origins, sources, and processes underscores notable co-benefits in control strategies. Quantitative assessments of measures outlined in the aforementioned action plans, particularly their impacts on reducing $PM_{2.5}$ concentrations and reducing CO_2 emissions, offer both theoretical insights and practical support for coordinated endeavors in climate change mitigation and air pollution abatement. As the potential of end-of-pipe control measures for emission reduction nears exhaustion, the imperative transition towards a low-carbon energy paradigm emerges as the linchpin for substantial air quality enhancement in China. This book further delves into case studies of three cities—Zhengzhou, Shijiazhuang, and Huzhou—to elucidate their distinctive attributes and lessons learned in achieving "dual attainments". Through these case analyses, common patterns are distilled to probe pathways and strategies for urban "dual attainments" within the context of air quality compliance.

Y. Bo et al., *Pathways for Reaching Air Quality Standard and Carbon Peak in Selected Cities in China*, Urban Sustainability, https://doi.org/10.1007/978-981-96-8315-4_1

1.1 China's Struggles in Attaining Air Quality Standards and Peaking Carbon Emissions

1.1.1 The Challenges Confronting China in the Pursuit of Air Quality Compliance

Since the commencement of reform and opening up policies, China has garnered worldwide attention for its exceptional economic progress. However, this rapid economic expansion has had adverse environmental implications. Significant urban agglomerations nationwide have encountered varying degrees of regional air pollution problems, sparking extensive public concern.

The endeavor to address air pollution in China commenced during the 1970s. Preceding the year 2000, it underwent successive phases of control, initially targeting suspended particulate matter emissions from industrial point sources (1970–1990) and subsequently, sulfur dioxide (SO_2) and suspended particulate matter emissions originating from coal combustion and industrial sources (1990–2000). Thereafter, it advanced into the control phase of addressing regional, integrated pollution stemming from multiple sources in 2000.

From 2001 to 2005, the overarching objective of China's comprehensive efforts to combat atmospheric pollution was to curtail national SO_2 emissions by 10% and reduce SO_2 emissions in areas designated for acid rain or SO_2 pollution control, commonly referred to as "two control zones", by 20%. This aimed to exert control over acid rain and SO_2 pollution at a nationwide level.

In the "National Total Emission Control Plan for Key Pollutants During the 11th Five-Year Plan Period", China incorporated SO_2 emissions in the national binding total emission control targets. This involved a priority focus on establishing desulfurization facilities in thermal power plants, with the goal of achieving a 10% reduction in national SO_2 emissions by 2010 compared to the levels in 2005.

During the 12th Five-Year Plan period, China further expanded its ambit by integrating nitrogen oxides (NOx) emissions into the national binding total emission control targets. The objectives entailed a 10% reduction in national NOx emissions and an 8% reduction in SO_2 emissions by 2015, compared to the levels in 2010.

In 2013, the State Council of China issued the "Air Pollution Prevention and Control Action Plan", hereinafter referred to as the "Ten Measures for Air". This landmark initiative delineated specific mandates for mitigating air pollution and established fine particulate matter ($PM_{2.5}$) concentrations as the controlling parameter. It marked a pivotal moment in China's endeavor to curb air pollution and signified the inaugural strategic action guided by environmental quality benchmarks. The "Ten Measures for Air" delineated principal tasks and measures, comprising 10 articles and 45 provisions. It also presented precise directives for enhancing air quality both nationally and in key regions and cities by 2017. Notably, this plan introduced innovative approaches to air pollution control, encompassing the restructuring of energy and industrial sectors and the implementation of major emission reduction projects. Additionally, it facilitated supplementary support in areas such as environmental law enforcement, environmental economic policies, and technological advancements. The confluence of these multifaceted measures has yielded favorable outcomes. Article 14 of the revised "Air Pollution Prevention and Control Law" in 2015 stipulates that city governments unable to meet the national air quality standards must promptly formulate remediation plans within specified timeframes and undertake effective measures to attain compliance within the durations stipulated by the State Council or provincial-level governments.

The "13th Five-Year Plan for Ecological and Environmental Protection", issued by the State Council, explicitly underscores the overarching requirement of "prioritizing environmental quality enhancement, and implementing air quality objective management along with time-bound compliance plans" [1]. In 2018, following the "Ten Measures for Air", the Ministry of Ecology and Environment unveiled the "Three-Year Action Plan for Winning the Blue Sky Defense Battle", hereinafter referred to as the "Three-Year Action Plan". This document further elucidated the fundamental approach, primary objectives, and central tasks for air pollution control efforts. The plan, centered on four pivotal elements, defined assignments, goals, and strategies for air pollution mitigation from 2018 to 2020. The "four key points" entail a primary emphasis on fine particulate matter as the chief pollutant, targeting key regions encompassing the Beijing-Tianjin-Hebei region and surrounding areas, the Yangtze River Delta region, and the Fenwei Plain, identifying specific seasons during autumn, winter, and early spring, and directing interventions toward crucial industries like steel, thermal power, and construction materials, as well as critical areas related to "scattered, non-compliant, and polluting" enterprises, bulk coal, diesel trucks, and dust control. The formulation of the "Three-Year Action Plan" aims to substantially curtail atmospheric pollutant emissions, visibly augment environmental air quality, and heighten public well-being under clear skies.

Since 2013, through an exceptionally arduous governance journey, China has achieved unparalleled accomplishments in its endeavors to mitigate air pollution. The 13th Five-Year Plan period stands out as the most triumphant phase in enhancing ecological and environmental quality, demonstrating remarkable progress and earning widespread acclaim from both the domestic populace and the international community. China has diligently fulfilled all stipulated objectives, outperforming the overarching benchmarks and quantifiable targets set forth during the 13th Five-Year Plan period. In 2020, the average concentration of $PM_{2.5}$ in 337 cities at the prefecture level and above across China stood at 33 $\mu g/m^3$, falling below the Grade II national standards for air quality (35 $\mu g/m^3$). In comparison to the 2015 average $PM_{2.5}$ concentration of 45 $\mu g/m^3$, this represented a substantial reduction of 26.7%, as depicted in Fig. 1.1. Over the period spanning from 2015 to 2020, the three-year moving average of $PM_{2.5}$ concentration in China consistently declined, manifesting a reduction of 57.1% in the incidence of heavily polluted days. This transformation resulted in an impressive 87% proportion of days characterized by excellent air quality [2].

Nonetheless, China's industrial, energy, and transportation structures have not yet undergone a foundational overhaul, and the carrying capacity of resources and the

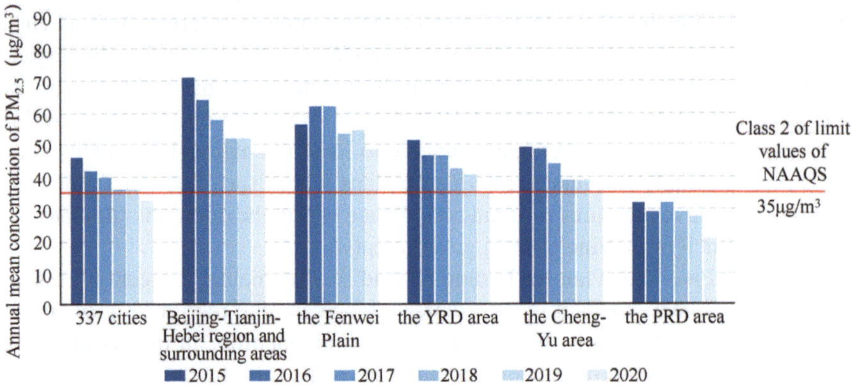

Fig. 1.1 Trends in average $PM_{2.5}$ concentrations across China and in key regions from 2015 to 2020

environment is either near or at the saturation point. China remains confronted with a situation characterized by substantial accumulation of ecological and environmental risks. Emissions of atmospheric pollutants, including SO_2, NOx, particulate matter, smoke, dust, and volatile organic compounds (VOCs), persist at a staggering level of ten million tons, significantly surpassing the ecological capacity. Furthermore, the burden of $PM_{2.5}$ pollution remains persistently elevated. In 2020, among the 337 cities at the prefecture level and above in China, 125 cities still fell short of meeting the standard for average $PM_{2.5}$ concentrations, and the problem of severe regional pollution during the autumn and winter months endured. Additionally, China's existing average standard for $PM_{2.5}$ air quality aligns merely with the first interim target established by the World Health Organization (WHO), highlighting a considerable disparity in comparison to the WHO guideline value. In 2020, the 90th percentile of the maximum daily 8-hour moving average ozone (O_3) concentration in Chinese cities averaged 136 μg/m^3, marking an 8.0% reduction from 2019 but an 11.1% increase compared to 2015, as depicted in Fig. 1.2. Concerning the three-year moving average, the mean O_3 concentration in China and key regions experienced a persistent upward trend during the period from 2015 to 2020. The national average O_3 concentration elevated from 123 μg/m^3 to 138 μg/m^3, with an accompanying increase in the proportion of cities surpassing the established standards, reaching 16.6%. O_3 concentration has become a pivotal factor influencing the number of days when urban air quality conforms to prescribed criteria.

In 2020, a significant 40% of cities across China remained unable to meet the prescribed air quality standards, with 37% of these cities experiencing $PM_{2.5}$ concentrations that exceeded the established limits. The need for coordinated efforts in managing $PM_{2.5}$ and O_3 has emerged as a paramount mission in the upcoming phase of air pollution control. This path remains challenging and extends into the distance. Concurrently, with the diminishing potential of end-of-pipe control measures and the narrowing scope for pollutant reduction, air pollution control is growing progressively intricate. The task of achieving substantial enhancements in air quality has

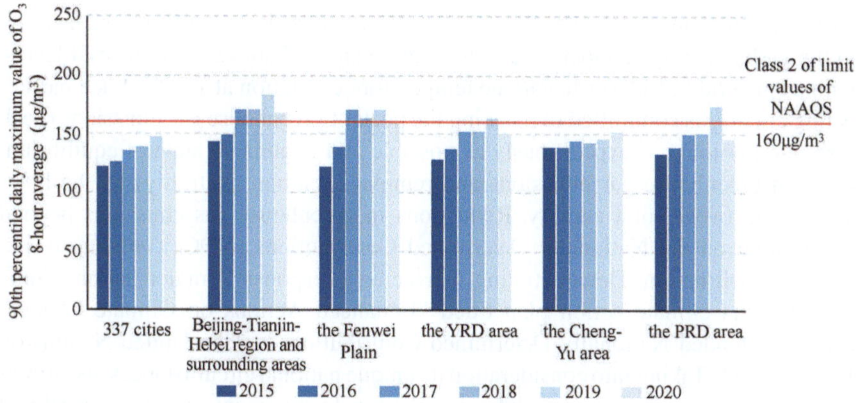

Fig. 1.2 Trends in the 90th percentile of the maximum daily 8-hour moving average O_3 concentration in China and key regions from 2015 to 2020

grown more formidable, underscoring the imperative to explore fresh avenues for air pollution management [2].

1.1.2 Hurdles on the Path to Attaining Peak Carbon Emissions in China

The Fifth Assessment Report of the Intergovernmental Panel on Climate Change (IPCC) emphatically underscores that human activities are likely the predominant catalyst for climate warming. Substantiated evidence demonstrates that the best estimate of total anthropogenic radiative forcing in 2011 surged by 43% when juxtaposed with the estimate in 2005 provided in the Fourth Assessment Report of the IPCC. Over the past half-century, a staggering tally of more than 11,000 weather, climate, and water-related disasters have unfolded across the globe, culminating in two million fatalities and economic losses reaching an astronomical \$3.6 trillion. The upshot of climate change is the heightened frequency, intensity, and gravity of extreme weather and climate events. The Sixth Assessment Report of the IPCC underscores that the global surface temperature during the period of 2011–2020 exceeded that of 1850–1900 by a notable 1.09 °C, with more pronounced increases over land (1.59 °C) than over the oceans (0.88 °C). As the global surface temperature continues its ascent, the coastal sea level in China has ascended by approximately 0.09 m over the past three decades, registering an average annual increase of 2.6 mm, exceeding the global average of 1.8 mm annually. The projection for China presages a continued upward trajectory in the annual average surface temperature.

In December 2015, nations from across the globe came together at the Paris Climate Change Conference and reached a unanimous accord, embracing the "Paris Agreement". The overarching objective of the "Paris Agreement" is to bolster the worldwide response to the peril of climate change. It seeks to constrain the increase in the global average temperature to no more than 2 °C above pre-industrial levels, with a concerted endeavor to cap the temperature escalation at 1.5 °C. Each participating country is committed to reaching the peak of greenhouse gas emissions (peak carbon emissions) as expeditiously as possible and attaining a state of equilibrium between greenhouse gas emissions and removals (carbon neutrality) by the latter half of the twenty-first century. Reductions in greenhouse gas emissions beyond 2020 hinge on the Nationally Determined Contributions (NDCs) voluntarily put forth by each nation. Demonstrating its role as a responsible major nation, China presented its climate action plan titled "Enhanced Actions on Climate Change: China's Intended Nationally Determined Contributions" to the United Nations on June 30, 2015. Taking into consideration its unique national circumstances, its current stage of development, its strategy for sustainable development, and its international obligations, China articulated a set of commitments to be achieved by 2030. These commitments encompass achieving the peak of carbon dioxide (CO_2) emissions, with aspirations to achieve this peak ahead of schedule, lowering CO_2 emissions per unit of gross domestic product (GDP) by 60%–65% relative to 2005 levels, elevating the share of non-fossil fuels in primary energy consumption to approximately 20%, and augmenting forest stock by approximately 4.5 billion cubic meters compared to 2005 levels. On September 3, 2016, the Standing Committee of the National People's Congress ratified China's accession to the "Paris Agreement". During the 75th session of the United Nations General Assembly in September 2020, the Chinese government declared its commitment to peaking CO_2 emissions prior to 2030 and attaining carbon neutrality by 2060. As of November 2020, 190 Parties have ratified the "Paris Agreement", and 188 Parties have submitted their NDCs [3]. In November 2021, during the 26th Conference of the Parties to the United Nations Framework Convention on Climate Change, a consensus was reached on fundamental issues concerning the operational aspects of the "Paris Agreement", heralding a new chapter in the collective global effort to comprehensively combat climate change.

Carbon emissions in China have evolved through three distinct phases. The period spanning from 1997 to 2001 was characterized by sluggish growth, with an annual average uptick in carbon emissions at a rate of 2.6%. In 2001, China's national carbon emissions stood at approximately 3.2 billion tons. The subsequent era, from 2002 to 2013, witnessed a swift expansion in carbon emissions. Driven by robust economic growth and the urbanization process, carbon emissions surged annually at an average rate of 9.6%. By 2005, China had overtaken the United States to become the world's foremost carbon-emitting nation. In 2013, China's per capita carbon emissions exceeded that of the European Union (6.8 tons), reaching 7.2 tons [4]. The period from 2014 to 2020 ushered in a phase of decelerated growth. Initially, there was a modest decline in carbon emissions, but in 2017, a growth trajectory reemerged. In 2020, China's national carbon emissions approximated 10 billion tons [5]. Since 2014, when the Chinese government initially proposed the goal of "striving to achieve the peak of CO_2 emissions as early as possible", a succession of national-level policy documents have been unveiled, elucidating explicit mandates concerning China's carbon emissions peak targets and the pathways to their realization. The

"Outline of the 13th Five-Year Plan for National Economic and Social Development of the People's Republic of China" (hereinafter referred to as the "Outline of the 13th Five-Year Plan") articulates the commitments to be attained by 2020. These include exerting effective control over carbon emissions in major carbon-intensive industries such as power, steel, building materials, and chemicals, while fostering low-carbon development across pivotal sectors like industry, energy, construction, transport, and more. Furthermore, the plan aims to bolster leading development regions in becoming early achievers in reaching their peak CO_2 emissions and to facilitate pilot programs for low-carbon development, exemplifying the establishment of near-zero carbon emissions zones. The "13th Five-Year Plan for Controlling Greenhouse Gas Emissions" provides additional benchmarks for 2020, including reducing carbon emissions per unit of GDP by 18% relative to 2015 levels, effectively managing total carbon emissions, and bolstering leading development regions in spearheading peak carbon emissions. The plan also endeavors to strive for early attainment of peak emissions in specific heavy and chemical industries and make tangible progress in the low-carbon transformation of energy systems, industrial systems, and consumption sectors. The majority of regions have been gradually advancing towards their respective carbon emissions peaks in line with national targets and overarching directives, based on their understanding of the peak targets and developmental stages.

In 2018, China successfully reduced its carbon intensity per unit of GDP by approximately 45.8% in comparison to 2015, surpassing the target of a 40%–45% reduction by 2020 ahead of schedule. However, data from several organizations, including the Carbon Dioxide Information Analysis Centre, the United Nations, and BP Amoco, reveal that due to factors such as an economic upswing, energy consumption in China displayed signs of resurgence in 2017. This led to a rebound in carbon emissions, breaking the trend of sustained reduction observed from 2013 to 2016. Peters et al. (2020) also noted that carbon emissions in China in 2018 experienced further growth compared to 2017. The reduction in carbon emissions achieved through a decrease in carbon intensity was offset by the emissions increase resulting from the recovery in heavy industrial production and augmented utilization of natural gas, as elucidated in Fig. 1.3. In the foreseeable future, China's overall energy demand is expected to continue its ascent, posing a formidable challenge to achieving the goal of peak CO_2 emissions before 2030.

China grapples with three significant challenges on its path toward achieving peak carbon emissions. The first challenge pertains to the high-carbon energy structure. In 2020, fossil fuel consumption constituted a staggering 84.1% of China's total energy consumption, with coal consumption alone accounting for 56.8% of fossil fuel usage. Both figures are the highest in the world, surpassing major energy-consuming nations like the United States. The second challenge centers on the high-carbon industrial structure. Energy-consuming industries known for their high carbon footprint and the difficulty of reducing emissions, such as coal, steel, petrochemicals, and cement, hold a substantial share within China's industrial landscape. The third challenge stems from China's status as the largest developing nation, where the urbanization process remains ongoing, accompanied by a surge in energy consumption in many regions. Global urbanization trends reveal that the advancement of urbanization and

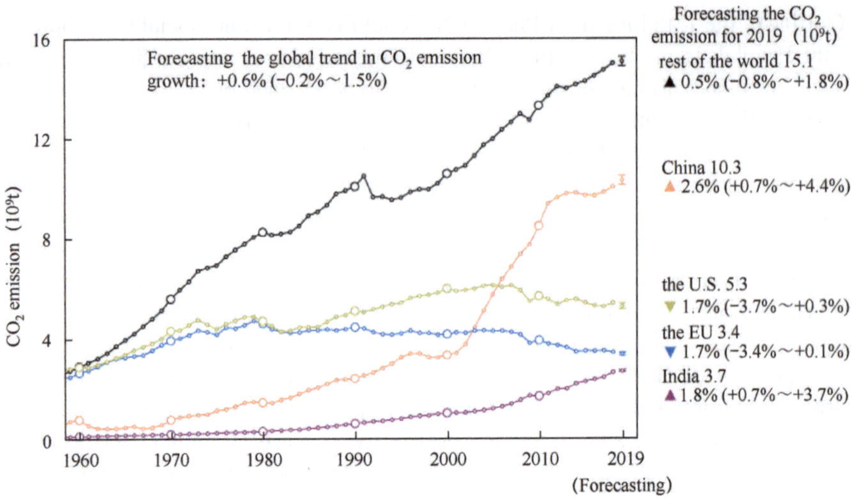

Fig. 1.3 Carbon dioxide emissions in China from 1960 to 2019 [6]

the enhancement of living standards often coincide with rapid growth in energy consumption. China's urbanization trajectory over the past three decades reaffirms this pattern. From 1990 to 2019, China's urbanization rate surged from 26.4% to 63.9%, per capita GDP escalated from 1700 yuan to 71,800 yuan, and per capita energy consumption rose from 0.87 tons of standard coal to 3.47 tons of standard coal. Presently, China's CO_2 emissions account for nearly 30% of the global total, which is approximately equivalent to the combined emissions of the United States and all European countries. Anticipating substantial energy demand in the near term, China faces a monumental challenge in adhering to its schedules for achieving peak carbon emissions and carbon neutrality. Furthermore, the processes of industrial upgrading and the establishment and promotion of non-fossil energy systems are time-intensive and capital-intensive endeavors. Given this context, China confronts considerable pressure in managing greenhouse gas emissions [7].

1.2 The Importance of Investigating a Coordinated Approach to Attaining Urban Air Quality Standards and Peaking Carbon Emissions

1.2.1 The Co-Benefits of Tackling Climate Change and Regulating Air Pollution

Previous research underscores the common origins, sources, and processes shared by greenhouse gases and atmospheric pollutants, all stemming from the combustion of fossil fuels like coal and oil [8–10]. The exploration of coordinated efforts to regulate greenhouse gases and atmospheric pollutants dates back to the 1990s when Ayres and Walter discussed the indirect benefits of greenhouse gas reduction, including the reduction in emissions of atmospheric pollutants and the corresponding health benefits [11]. In 2001, the "Third Assessment Report of the IPCC" introduced the concept of co-benefits derived from simultaneously implementing policies for diverse objectives, including climate change mitigation. It acknowledged that many policies aimed at greenhouse gas mitigation also serve other, equally significant objectives, such as curbing atmospheric pollutants. The "Fourth Assessment Report of the IPCC" further emphasized that the notion of co-benefits is often linked to "no-regret" policies. This association arises from the observation that cost-benefit analyses of emissions reduction in numerous projects and industries have revealed that greenhouse gas mitigation policies may exhibit the potential for negative costs, meaning the co-benefits derived from implementing these policies outweigh their implementation expenses. Consequently, such emissions reduction policies with negative costs are frequently termed "no-regret" policies. These co-benefits not only enhance public health but also have implications for agricultural production and natural ecosystems. Coordinated policies aimed at curbing air pollution and mitigating climate change hold the promise of substantial cost reductions when compared to standalone policies [12].

The "Fifth Assessment Report of the IPCC" made a distinction between positive co-benefits and negative co-benefits (unintended adverse side effects). It delved into the technical, economic, and institutional prerequisites for greenhouse gas mitigation strategies, along with their potential positive co-benefits or adverse side effects [13]. The "IPCC Special Report on Global Warming of 1.5°C", issued in 2018, placed greater emphasis on the concept of co-benefits, specifically positive impacts. It defined co-benefits as the positive outcomes that policies or measures targeting a specific goal may have on other objectives, thereby augmenting overall societal or environmental benefits. The evaluation of co-benefits is often contingent on uncertain variables and relies on the specific local conditions and policy implementation context [7, 14]. In line with these principles, the Working Group II report of the "Sixth Assessment Report of the IPCC", titled "Climate Change 2022: Impacts, Adaptation, and Vulnerability", published in February 2022, suggested that comprehensive, effective, and innovative mitigation measures can leverage co-benefits to alleviate

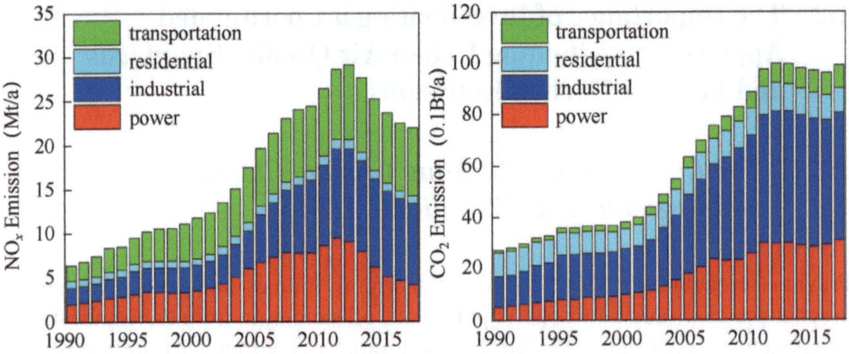

Fig. 1.4 Sectoral emissions of NO_x and CO_2 in China from 1990 to 2017 (Source: Multi-resolution Emission Inventory for China, Tsinghua University)

trade-offs between adaptation and mitigation, consequently fostering sustainable development [15].

Global climate change and local air pollution concerns are chiefly propelled by comparable energy production and consumption patterns, which align their working directions closely. Presently, numerous domestic and international studies have conducted quantitative assessments of the extent of co-benefits arising from the reduction of greenhouse gases and atmospheric pollutant emissions. Figure 1.4 illustrates the sectoral emissions of NOx and CO_2 in China from 1990 to 2017. It is evident that they manifest analogous trends in variation and sectoral composition, affirming the viability of coordinated governance policies.

Since 2013, the consecutive implementation of the "Ten Measures for Air" and the "Three-Year Action Plan" has introduced a series of structural control measures within the realm of air pollution prevention and control. These measures encompass actions like the renovation of coal-fired boilers, the elimination of outdated production capacity, the adoption of clean heating in northern regions, and the restructuring of transportation systems. These actions have generated positive co-benefits in terms of reducing CO_2 emissions. Driven by the vigorous impetus provided by these initiatives, the industrial sector achieved synchronized reductions in both CO_2 emissions and atmospheric pollutants from 2015 to 2020, resulting in a 6% decrease in CO_2 emissions. Meanwhile, in the civil sector, reductions in major atmospheric pollutant emissions have occurred alongside relatively stable CO_2 emissions. However, in the electricity, heating, and transportation sectors, declines in major atmospheric pollutant emissions were coupled with continued increases in CO_2 emissions. To elaborate further, the reduction of atmospheric pollutants in the electricity and heating sectors was primarily reliant on end-of-pipe control measures, making it challenging to achieve concurrent reductions in CO_2 emissions. In the transportation sector, the decline in atmospheric pollutant emissions was primarily a consequence of end-of-pipe measures, such as the retirement of older vehicles and the enhancement of emission standards. Nevertheless, the rapid proliferation of motor vehicles in China offset the emission reduction benefits derived from policy implementation. From 2015 to 2020, CO_2 emissions from the transportation sector rose by 14% [16].

According to an evaluation conducted by the Chinese Academy of Engineering, both the "Ten Measures for Air" and the "Three-Year Action Plan" have been successful in reducing the national $PM_{2.5}$ concentration. However, when comparing the periods of implementing these two plans, the improvement in $PM_{2.5}$ concentration during the "Three-Year Action Plan" implementation has shown a diminishing rate of progress. Furthermore, there have been significant changes in the ranking of contributions from various measures to the improvement of $PM_{2.5}$ concentration.

Notably, the contribution of industrial upgrading and renovation, as well as the transition to cleaner energy sources in the civil sector, has increased and emerged as the most substantial factors, accounting for 21% and 18% of the overall reduction in $PM_{2.5}$ concentration, respectively, as illustrated in Fig. 1.5.

Throughout the implementation of the "Ten Measures for Air", there was a cumulative reduction of 920 million tons of CO_2 emissions. However, during the execution of the "Three-Year Action Plan", the cumulative reduction in CO_2 emissions amounted to 490 million tons. In comparison to the period when the "Ten Measures for Air" were enforced, the contributions to coordinated CO_2 emissions reduction from actions such as eliminating excess capacity, upgrading coal-fired boilers, and controlling emissions from mobile sources not only narrowed but also exhibited reduced reduction magnitudes. The emissions reduction effects of cleaning and rectifying "scattered, non-compliant, and polluting" enterprises have become more

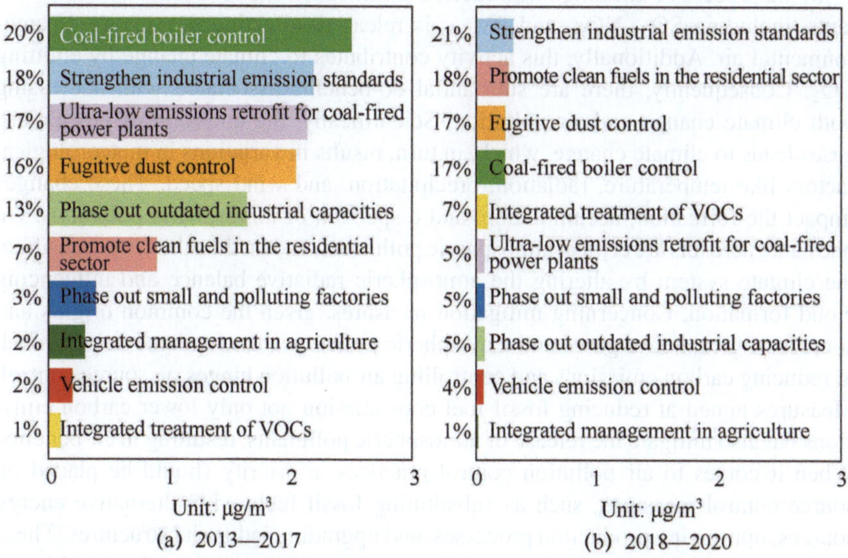

Fig. 1.5 Contribution of measures from the "Ten Measures for Air" and the "Three-Year Action Plan" to the enhancement of $PM_{2.5}$ concentrations

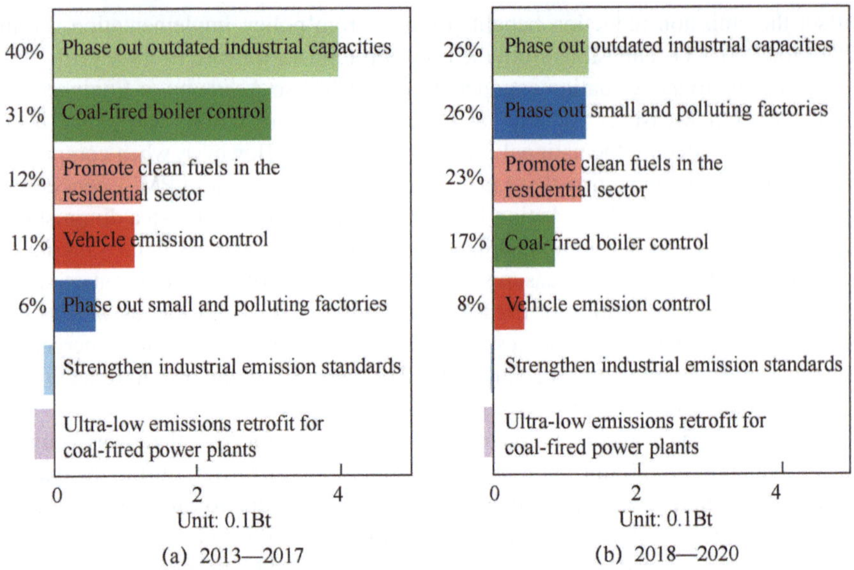

Fig. 1.6 Similar to Fig. 1.5, but for coordinated CO_2 emissions reduction

prominent, primarily due to the substantial progress made in governing kilns in the brick and lime industry, as depicted in Fig. 1.6.

In the process of utilizing fossil fuels, a significant volume of atmospheric pollutants, including SO_2, NOx, and $PM_{2.5}$, is released, affecting the quality of environmental air. Additionally, this activity contributes to climate change by emitting CO_2. Consequently, there are substantial co-benefits associated with addressing both climate change and air pollution. Scientifically, the emission of greenhouse gases leads to climate change, which, in turn, results in variations in meteorological factors like temperature, radiation, precipitation, and wind speed. These changes impact the formation, accumulation, and dispersion of atmospheric pollutants. On one hand, aerosols are critical atmospheric pollutants, and on the other, they can affect the climate system by altering the atmospheric radiative balance and influencing cloud formation. Concerning mitigation measures, given the common origins and sources of greenhouse gases and atmospheric pollutants, the fundamental approach to reducing carbon emissions and controlling air pollution hinges on source control. Measures aimed at reducing fossil fuel consumption not only lower carbon emissions but also mitigate the release of atmospheric pollutants, resulting in co-benefits. When it comes to air pollution control practices, a priority should be placed on source control measures, such as substituting fossil fuels with alternative energy sources, optimizing production processes, and upgrading industrial structures. These measures not only reduce emissions of atmospheric pollutants but also yield benefits in terms of carbon emissions reduction. The potential for reducing atmospheric pollutants through adjustments in China's energy and transportation sectors is an area that warrants further exploration. The next step should involve actively promoting source reduction measures with the goal of synergizing the reduction of pollutant and carbon emissions.

1.2.2 The Analysis of the Potential for Coordinated Reduction of Greenhouse Gases and Pollutants

Relevant studies suggest that if China simultaneously intensifies efforts to reduce carbon emissions and enhance source control, alongside continuous pollution control measures in non-electric industries, diesel engines, and key VOCs industries, it is feasible to achieve peak carbon emissions by 2030. This would be accompanied by a reduction in national VOCs emissions by 29% and a 51% decrease in SO_2 emissions compared to 2015 [16]. Additionally, the annual average concentration of $PM_{2.5}$ in most regions could meet the current Grade II national standards for air quality, which is set at 35 μg/m^3. By increasing the proportion of renewable energy generation and accelerating the shift toward energy-saving and electrification in the end-use system, China could achieve an early peak in CO_2 emissions by 2025. Furthermore, this approach could reduce the national $PM_{2.5}$ exposure level to around 23 μg/m^3 by 2030, as depicted in Fig. 1.7 [17]. Beyond 2030, as the emission reduction potential of end-of-pipe control measures becomes largely exhausted, the deep transition to a low-carbon energy system, aligned with the goal of carbon neutrality, will be the driving force for the sustained and profound improvement of air quality in China. Under the carbon neutrality scenario, China is expected to have completed its transition to a low-carbon energy system by 2060, resulting in a reduction of national total carbon emissions by over 90% from current levels. Moreover, compared to 2015 levels, national VOCs and SO_2 emissions are projected to decrease by approximately 65% and 94% [16], respectively. The annual average $PM_{2.5}$ exposure level for the population is anticipated to reach around 8 μg/m^3, fundamentally resolving air pollution issues [18]. Consequently, aligning climate goals with CO_2 emissions reduction and coordinated reduction of major pollutants is the inevitable path for China's medium- and long-term climate and environmental governance.

It is important to recognize that there exist certain inconsistencies in China's climate and environmental governance efforts. On one hand, China has made significant strides in reducing SO_2 emissions in recent years, and further decreases in SO_2 emissions are expected as a result of climate and environmental governance policies. However, this reduction may potentially contribute to heightened regional radiative forcing and accentuate global warming trends, consequently leading to adverse climate effects. On the other hand, black carbon particles play a critical role in atmospheric warming, and thus, efforts to reduce black carbon emissions can help alleviate climate warming. However, the sources of black carbon and cooling agents in the atmosphere, such as sulfates and organic carbon, are closely intertwined, underscoring the complexity of region-specific emissions reductions within

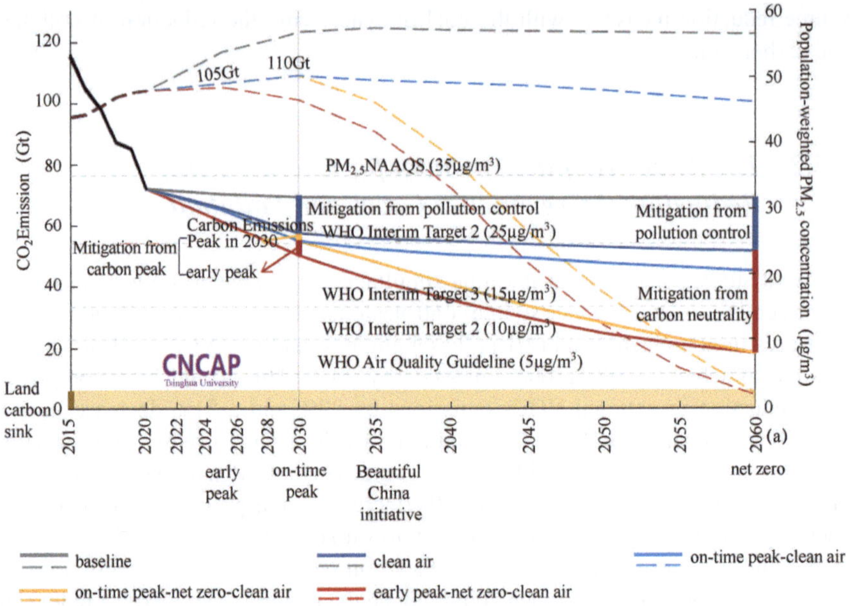

Fig. 1.7 A coordinated approach to achieving carbon neutrality and improved air quality in China (2015–2060) [18]

the framework of coordinated environmental and climate governance. Moreover, the installation of end-of-pipe emissions control devices can lead to a minor increase in CO_2 emissions, including the CO_2 emissions indirectly generated by additional electricity consumption and CO_2 released from chemical reactions involved in sulfur and nitrogen absorption. Nevertheless, this incremental increase is generally modest. For instance, the extra electricity generation resulting from the implementation of wet flue gas desulfurization systems in power plants accounts for approximately 2%–3% of total electricity generation [19]. Hence, when devising emission reduction strategies, it is imperative to take a comprehensive approach, considering their impact on both air quality and climate, to effectively achieve harmonized governance that addresses both climate change and environmental pollution.

1.2.3 The Need for Coordinated Strategies to Simultaneously Attain Urban Air Quality Standards and Peak Carbon Emissions

In China, the urban population accounts for 50% of the total, consuming more than 80% of the nation's energy and contributing to over 80% of carbon emissions as well as more than 60% of atmospheric pollutant emissions. Consequently, cities hold the key to achieving coordinated reductions in emissions [20]. Notably, since the institutional reform of the State Council in 2018, the responsibility for addressing climate change has shifted from the National Development and Reform Commission to the Ministry of Ecology and Environment. Furthermore, the ecological and environmental departments have taken on the climate change responsibilities of all 31 provinces, autonomous regions, and municipalities directly under the central government. This transition poses a novel challenge for regulatory authorities, who must determine how to attain coordinated governance of greenhouse gases and atmospheric pollutants to reap the co-benefits of climate change mitigation and air quality enhancement. Cities, as the fundamental units for policy implementation, play a pivotal role in addressing this challenge.

In September 2020, President Xi Jinping delivered a significant proclamation at the 75th session of the United Nations General Assembly, committing China to strive for a peak in CO_2 emissions before 2030 and pursue carbon neutrality by 2060. These new climate objectives have invigorated the ongoing efforts to enhance air quality in China. The country now grapples with the dual challenge of intensifying its anti-pollution measures while advancing toward the goals of "peak carbon emissions and carbon neutrality". To tackle this, the focus should be on reinforcing the governance and control of emission sources, prioritizing energy efficiency and conservation, transitioning to low-carbon and clean energy sources, and rationalizing energy allocation. This strategic approach will promote the synergetic reduction of pollutant and carbon emissions, thus amplifying the co-benefits. Currently, there are still challenges in synergizing the reduction of pollutant and carbon emissions. These include the lack of comprehensive top-level planning and policy arrangements, incomplete greenhouse gas emission data in various regions, and a shortage of personnel and capabilities to coordinate efforts for pollutant and carbon reduction. Comparatively, the current urban air quality management system is more well-developed with clear objectives. However, most climate change mitigation endeavors remain at the macro level. Cities need to fully recognize the pivotal role and significance of synergetic reduction of pollutant and carbon emissions. They should regard it as the overarching strategy for propelling a comprehensive, green transformation in economic and social development. By adopting this approach, we can effectively drive environmental quality improvement, achieve the goal of peaking carbon emissions by 2030, and attain carbon neutrality by 2060, ultimately fostering high-quality economic development.

As of the end of April 2020, 58 cities have formulated and published plans aimed at attaining air quality standards within specific timeframes, while 82 cities have successfully completed low-carbon city implementation plans. Several studies have conducted coordinated emission reduction analyses for individual cities. However, there remains a paucity of research examining coordinated emission reduction effects of cities and their future coordinated emission reduction trajectories on a national scale (refer to Table 1.1). Within China, cities exhibit varying levels of development, economic diversification, and distinct differences in emission sources and potential for emission reduction [21]. It is, therefore, of paramount importance to establish customized coordinated emission reduction strategies, aligned with the unique

Table 1.1 Illustrative Instances of coordinated emission reduction research

Study Title	City	Assessment of Historical Coordinated Emission Reduction Effects	Analysis of Future Coordinated Emission Reduction Trajectories
Chen et al., 2006	Shanghai		√
Li Liping et al., 2010	Panzhihua	√	
Liu et al., 2013	Beijing		√
Feng Xiangzhao et al., 2018	Chongqing	√	
Achieving Triple Goals of Carbon Emission Peaking, Air Quality Standard Attainment, and Economic Prosperity at the City Level: the Shenzhen Case, 2019	Shenzhen	√	√
Xing Youkai et al., 2020	Tangshan	√	
Assessment Report on the Synergistic Management of Carbon Dioxide and Air Pollution in Chinese Cities, 2020	335 prefecture-level cities	√	

characteristics of individual cities. This approach will significantly contribute to the collective achievement of improved air quality and reduced greenhouse gas emissions goals across cities in the future.

Given these considerations, this book puts forth a methodological framework for the synchronized examination of accomplishing air quality standards and peak carbon emissions (hereafter referred to as "dual attainments"). Within the pages of this book, three cities displaying distinctive traits—Zhengzhou, Shijiazhuang, and Huzhou—are chosen to illuminate the individual attributes and lessons learned by these cities in attaining "dual attainments". Subsequently, from these personalized experiences, common patterns are distilled to investigate the pathways and strategies for urban "dual attainments" driven by air quality objectives. The aim of this book is to offer research materials and policy formulation support for urban-level endeavors towards "dual attainments" within the context of national and regional institutional reforms.

References

1. Tang, Q., Chen, X.J., Huang, Y., et al. Study on Roadmap for City Air Quality Compliance: A Case Study of Wuhan[J]. Environmental Pollution & Control (in Chinese), 2019,41(12):1512-1516.
2. Working Group of Synergetic Roadmap on Carbon Neutrality and Clean Air for China, Synergetic Roadmap on Carbon Neutrality and Clean Air for China (in Chinese), 2021[R]. Beijing: China Clean Air Policy Partnership, 2021. https://www.efchina.org/Reports-zh/report-cemp-20211020-zh.

3. IRENA. Renewable Energy and Climate Pledges: Five Years after the Paris Agreement[R]. 2020-12.
4. Quéré, C.L., Moriarty, R., Andrew, R.M., et al. Global Carbon Budget 2014[J]. Earth System Science Data, 2015, 7(1): 47-85.
5. Shan, Y., Guan, D., Zheng, H., et al. China CO2 Emission Accounts 1997-2015[J]. Scientific Data, 2018, 5(1): 170-201.
6. Peters, G.P., Andrew, R.M., Canadell, J.G., et al. Carbon Dioxide Emissions Continue to Grow Amidst Slowly Emerging Climate Policies[J]. Nature Climate Change, 2020, 10: 3-6.
7. Wang, C., Deng, H.M., Guo, K.D., et al. Research Progress on Cooperative Governance of Greenhouse Gases and Air Pollutants[J]. Chinese Journal of Environmental Management (in Chinese), 2020,4:5-12.
8. Xue, J., Luo, H., Lv, L.H., et al. Emission Characteristics and Correlation of Main Air Pollutant and Greenhouse Gas in China[J]. Resources Science (in Chinese), 2012(8):1452-1460.
9. He, J.Y. Study on the Cooperative Control of Greenhouse Gases and Local Atmospheric Pollutants (in Chinese) [D]. Taiyuan: Shanxi University, 2011.
10. Clean Air Alliance of China. CAAC Clean Air Management Report 2016 (in Chinese) [R]. Beijing: Clean Air Alliance of China, 2016. https://www.efchina.org/Reports-zh/report-cemp-20161020-1-zh/view.
11. Ayres, R.U., Walter, J. The Greenhouse Effect: Damages, Costs and Abatement[J]. Environmental and Resource Economics, 1991, 1(3): 237-270.
12. IPCC. Climate Change 2007: Mitigation-Contribution of Working Group III to the fourth Assessment Report of the Intergovernmental Panel on Climate Change[M]. Cambridge: Cambridge University Press, 2007.
13. IPCC. Climate Change 2014: Mitigation of Climate Change-Contribution of Working Group III to the fifth Assessment Report of the Intergovernmental Panel on Climate Change[M]. Cambridge: Cambridge University Press, 2014.
14. IPCC. Global Warming of 1.5°C[Z]. An IPCC Special Report on the Impacts of Global Warming of 1.5°C above Pre-industrial Levels and Related Global Greenhouse Gas Emission Pathways, in the Context of Strengthening the Global Response to the Threat of Climate Change, Sustainable Development, and Efforts to Eradicate Poverty. 2018.
15. IPCC. Climate Change 2022: Impacts, Adaptation and Vulnerability-Contribution of Working Group II to the Sixth Assessment Report of the Intergovernmental Panel on Climate Change[M]. Cambridge: Cambridge University Press, 2022.
16. Tong, D., Cheng, J., Liu, Y., et al. Dynamic Projection of Anthropogenic Emissions in China: Methodology and 2015-2050 Emission Pathways under a Range of Socio-Economic, Climate Policy, and Pollution Control Scenarios[J]. Atmospheric Chemistry and Physics, 2020, 20(9): 5729-5757.
17. Tong, D., Geng, G., Zhang, Q., et al. Health Co-Benefits of Climate Change Mitigation Depend on Strategic Power Plant Retirements and Pollution Controls[J]. Nature Climate Change, 2021, 11(12):1077-1083.
18. Cheng, J., Tong, D., Zhang, Q., et al. Pathways of China's PM2.5 Air Quality 2015-2060 in the Context of Carbon Neutrality[J]. National Science Review, 2021, 8(12): 12.
19. Lu, Z., Huang, L., Liu, J., et al. Carbon Dioxide Mitigation Co-Benefit Analysis of Energy-Related Measures in the Air Pollution Prevention and Control Action Plan in the Jing-Jin-Ji region of China[J]. Resources, Conservation& Recycling, 2019, X, 1: 100006.
20. Zheng, B., Zhang, Q., Steven, J.D., et al. Infrastructure Shapes Differences in the Carbon Intensities of Chinese Cities[J]. Environmental Science & Technology, 2018, 52(10): 6032-6041.
21. Shan, Y., Guan, D., Hubacek, K., et al. City-Level Climate Change Mitigation in China[J]. Science Advances, 2018, 4(6): eaaq0390.

Chapter 2
Materials and Methods

Abstract This chapter furnishes an intricate introduction to the emission inventory, model simulation, measure assessment, and the methodology for analyzing "dual attainments" within the research framework. It undertakes an in-depth exploration of the evolution, categorization, data acquisition, and computational techniques employed in emission inventories. Moreover, it outlines the methodology for compiling integrated emission inventories encompassing atmospheric pollutants and greenhouse gases. Central to this discussion is the introduction of a hierarchical emission source classification system, organized into ten anthropogenic emission categories, and further subdivided into sector/industry, fuel/product, combustion/process technology, and end-of-pipe control technology levels. This chapter establishes a dynamic emission characterization model to facilitate a nuanced understanding of emissions dynamics. Leveraging the WRF-CMAQ model, simulations of atmospheric physical and chemical processes are conducted. These simulations, in conjunction with the integrated emission inventory, enable quantitative analysis of emission reduction quantities and the contributions of various mitigation measures, with the goal of supporting the formulation of plans for air quality compliance. Furthermore, this chapter employs the Long Range Energy Alternatives Planning (LEAP) model to scrutinize pathways towards urban peak carbon emissions. It elucidates the foundational assumptions, requisites, and transformations inherent in the LEAP model while constructing a technical roadmap for analyzing trajectories towards peak carbon emissions driven by urban air quality compliance.

2.1 Emission Inventory

2.1.1 Overview of Emission Inventory

A compendium denoted as an atmospheric pollutant emission inventory (hereinafter designated as the "emission inventory") encompasses the aggregation of quantified pollutant volumes discharged into the atmosphere from diverse emission sources

Y. Bo et al., *Pathways for Reaching Air Quality Standard and Carbon Peak in Selected Cities in China*, Urban Sustainability, https://doi.org/10.1007/978-981-96-8315-4_2

within a specified temporal domain and spatial demarcation. This compilation delineates variations in land-air and sea-air fluxes, elucidating the mechanisms governing the release of trace atmospheric constituents from disparate mediums. Furthermore, it constitutes an indispensable foundational dataset for scrutinizing global biogeochemical cycles, encompassing but not limited to carbon, nitrogen, sulfur, and iron cycles. It facilitates the analysis of shifts in atmospheric components and provides a framework for interpreting observational data related to atmospheric phenomena. Consequently, the emission inventory assumes paramount significance in the investigation of atmospheric reaction processes and the comprehension of mechanisms underpinning pollution formation.

In the realm of air quality management, the emission inventory stands as a pivotal instrument for discerning pollution sources, devising emission reduction strategies, and appraising the efficacy of control measures. Exemplary emission inventories form the bedrock of air quality management practices globally. Taking the United States (U.S.) and Europe as illustrative cases, the U.S. enacted the "Clean Air Act" in 1963, followed by the "Clean Air Act Amendments" in 1970. Under the auspices of the U.S. Environmental Protection Agency, a methodological framework for compiling emission inventories, inclusive of comprehensive databases, emission source processing models, and validation systems, was systematically established. This concerted effort culminated in the development of the National Emission Inventory (NEI). Concurrently, Europe embarked on its emission inventory development in the 1980s, adopting a unified and publicly shared methodology across European nations. Notable European emission inventories include CORe INventory AIR Emissions (CORINAIR) and the European Monitoring and Evaluation Programme (EMEP), encompassing more than 30 European countries and cataloging over 200 major anthropogenic emission sources. This approach ensures the scientific rigor and efficacy of pollution control planning. Advancements in atmospheric chemistry, steered by the International Global Atmospheric Chemistry (IGAC) program, reveal that atmospheric pollutants exert far-reaching impacts on atmospheric constituents through intricate physical and chemical transformations. This engenders either unidirectional influence or bidirectional feedback loops that resonate globally, impacting climate, ecosystems, and public health. Such repercussions ultimately mold economic and social development strategies aimed at regulating and managing pollutant emissions. In this iterative process, the recognition and comprehensive understanding of atmospheric pollutant emissions serve as the foundational bedrock and inaugural point for research endeavors, as well as the focal point for policy management.

Diverse research perspectives and calculation scales afford varied classifications for emission inventories.

Emission inventories, contingent upon the origin of emissions, undergo classification into anthropogenic source and natural source emission inventories. Anthropogenic source emissions emanate from human activities, encompassing phenomena such as fuel combustion, industrial production, nitrogen fertilizer application, and the utilization of paints. These activities contribute to the release of atmospheric pollutants. In contrast, natural source emissions arise from inherent natural processes, such as volcanic eruptions, high-altitude lightning, and emissions from vegetation. Anthropogenic source emissions exhibit distinctive characteristics in terms of intensity and spatiotemporal extent when juxtaposed with natural source emissions. The former, characterized by continuity and concentration, assume a preeminent role in the discharge of significant atmospheric pollutants, including SO_2, NO_x, and $PM_{2.5}$, as well as greenhouse gases such as CO_2, N_2O, and CH_4, as opposed to the sporadic and diverse nature of emissions from natural sources.

Depending on the scope of research, emission inventories are systematically classified into global, regional, and local categories. Global emission inventories, typically formulated at the national level, rely on macroeconomic sectoral statistics for the computation of emissions associated with major atmospheric pollutants. Generally, these global inventories encompass a narrower array of emission sources. Regional emission inventories, conversely, are developed at the national or regional level, with calculations conducted at the provincial or state level. These inventories delineate emission sources with specificity down to the industry level. Local emission inventories, situated at the city level, provide highly detailed specifications regarding emission source types and calculation scales. In recent years, a noteworthy technical trend within the field involves amalgamating precision regional or local emission inventories from diverse areas to construct expansive, high-resolution emission inventories. Emission inventories resulting from this technical approach not only span global or hemispheric scales but also attain data accuracy comparable to regional or local emission inventories in key areas. This emerging approach currently represents a focal point of research interest.

Figure 2.1 encapsulates the contemporary landscape and significant strides in emission inventory technology research. This technological domain predominantly addresses the challenge of constructing emissions inventories ab initio, entailing the formulation of emission characterization models grounded in pertinent emission source classification systems, the modeling and parameterization of key influencing factors, and the ultimate establishment of comprehensive emission inventories. Within this research realm, various standardized technical approaches have emerged, encompassing inventory techniques such as dynamic process-based emission characterization, uncertainty estimation, historical trend reconstruction, and future emission projections. Additionally, these approaches include techniques integrated with atmospheric chemistry models, incorporating emission source processing models and high-resolution inventory techniques. The application of these common technical approaches across local and global/regional scales encounters diverse data foundations and technical challenges, resulting in attributes of multi-source data assimilation for local emission inventory techniques and features of large-scale, long-term, and multi-perspective considerations in global/regional emission inventory techniques [1].

Figure 2.2 elucidates the fundamental technical facets of emission feature analysis. This strand of research predominantly concentrates on three key dimensions: emission composition, emission time-varying characteristics, and emission spatial distribution. These dimensions align with the assessment of emission quantity, temporal variations in emissions, and the spatial distribution of emissions. The analysis delves

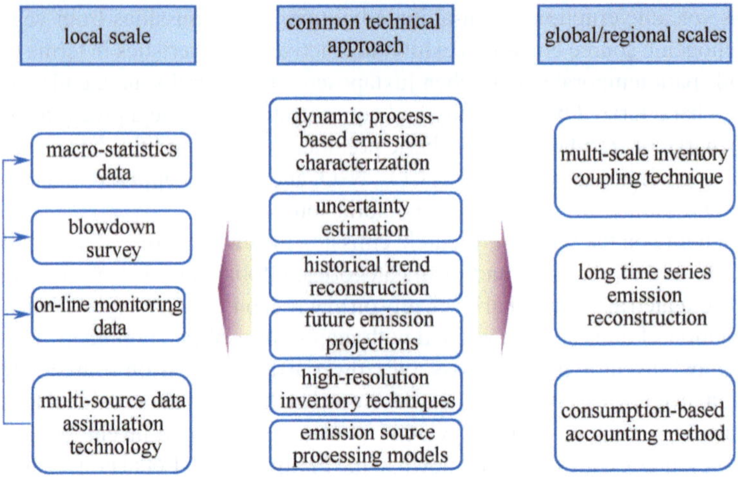

Fig. 2.1 Emission inventory techniques

into the attributes of emission intensity and the spatiotemporal extent of emissions on a comprehensive scale.

Studies on emission composition are centered on the meticulous analysis of aggregate emissions, uncertainty ranges, and the distribution of emission sectors. These investigations aim to enhance the precision of emission inventories through comparative analyses of distinct emission inventories. The scrutiny of emission time-varying characteristics focuses on discerning interannual and monthly variations in emissions, examining the fundamental drivers and influencing factors behind emission fluctuations. Additionally, it incorporates an examination of hourly variations to capture the dynamic activity patterns inherent in emission source intensity. Emission

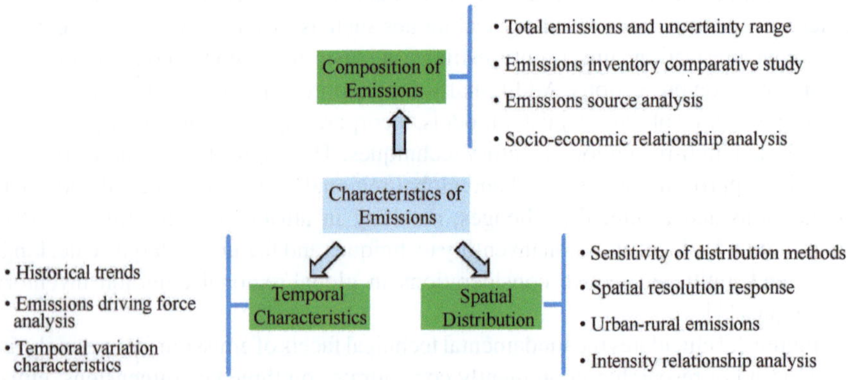

Fig. 2.2 Primary technical characteristics of emission feature analysis

spatial distribution analysis delves into the morphological aspects of emission distribution, encompassing the study of emission spatialization methods and the identification of emission spatial distribution patterns. It is noteworthy that, with the advancement of high-resolution emission inventory techniques, research on emission spatial distribution has progressively introduced novel perspectives and achievements. For example, leveraging correlation analysis between high-resolution emission intensity distribution and urban development morphology enables an exploration of urban morphological characteristics contributing to the emergence of low-emission areas. This, in turn, provides valuable support for the formulation of sustainable urban development strategies.

2.1.2 Urban Emission Inventory Compilation

The establishment of unified greenhouse gas and air pollutant emission inventories at the urban scale within a cohesive methodological framework serves as the cornerstone for navigating pathways towards urban "dual attainments". Traditionally, the compilation of greenhouse gas and atmospheric pollutant emission inventories has occurred in isolation, leading to challenges such as divergent accounting boundaries, statistical methodologies, and categorizations of emission sources. This compartmentalization has impeded the identification of key areas for emission reduction and the calculation of coordinated reduction potentials. Hence, when undertaking the compilation of urban-scale greenhouse gas and atmospheric pollutant emission inventories, the utilization of a unified emission source classification system and emission characterization techniques is imperative.

Tsinghua University has pioneered the establishment of a technical system for compiling urban atmospheric pollutant emission inventories, comprising the "Manual for Compilation of Urban Atmospheric Pollutant Emission Inventory", "Technical Manual for Compilation of Urban Atmospheric Pollutant Emission Inventory", "Technical Specifications for Activity Level Surveys of Atmospheric Pollutant Emission Inventory", and the "Atmospheric Pollutant Emission Inventory Compilation and Analysis System".

- The "Manual for Compilation of Urban Atmospheric Pollutant Emission Inventory" provides a detailed outline of the procedural steps involved in compiling urban atmospheric pollutant emission inventories. It functions as a guiding document for key regional cities across the nation, assisting them in the effective execution of emission inventory compilation.
- The "Technical Manual for Compilation of Urban Atmospheric Pollutant Emission Inventory" introduces a four-level emission source classification system, anchored in ten categories of anthropogenic emissions. This manual encompasses 3299 emission source categories and associated accounting methods for nine distinct air pollutants. Furthermore, it supplies temporal profiles for various emission sources, facilitating dynamic emission accounting.

- The "Technical Specifications for Activity Level Surveys of Atmospheric Pollutant Emission Inventory" delineates the essential components for conducting a comprehensive survey of emission sources, accompanied by in-depth technical explanations.
- The conception and development of the "Atmospheric Pollutant Emission Inventory Compilation and Analysis System" are rooted in the previously delineated workflow, technical methodologies, and survey specifications. This sophisticated system seamlessly incorporates hierarchical classification and coding systems for emission sources, techniques for compiling urban atmospheric pollutant emission inventories, and a nexus of connections, encompassing activity-level data collection, emission factor acquisition, dynamic emission inventory approaches, and emission spatial allocation. A user-friendly interface empowers this system to conduct background calculations on input data, generate emission quantities, and streamline the processes of emission inventory verification and data analysis.

This publication leverages the previously outlined technical system methodologies to construct an urban-scale emission inventory encompassing both greenhouse gases and atmospheric pollutants. This comprehensive emission inventory encapsulates all anthropogenic emission sources within the administrative jurisdictional boundaries, encompassing gaseous and particulate pollutants like SO_2, NO_x, VOCs, CO, NH_3, $PM_{2.5}$, black carbon (BC), and organic carbon (OC), in addition to the pivotal greenhouse gas, CO_2.

From the standpoint of achieving air quality standards, the compilation of the emission inventory solely necessitates consideration of atmospheric pollutant emissions stemming from primary energy consumption and industrial processes within the city boundaries. There is no imperative to account for atmospheric pollutant emissions originating from secondary energy sources, such as electricity and heat consumption, and upstream activities related to the production of goods. The formula for calculating atmospheric pollutant emissions is articulated as follows:

$$E_i = \sum_j \sum_k \sum_m \sum_n A_{i,j,k} X_{i,j,k,m} \mathrm{EF}_{\mathrm{RAW},i,j,k,m} C_{i,j,k,m,n} (1 - \eta_n) \tag{2.1}$$

where i represents the city, j denotes the emission sector/industry, k represents the fuel/product, m denotes the combustion/production technology, n is the end-of-pipe control technology, A denotes the activity level, such as fuel consumption or product output, X is the proportion of a specific combustion/production activity level to the total activity level, $\mathrm{EF}_{\mathrm{RAW}}$ represents the emission factor for atmospheric pollutants without end-of-pipe control technology, C is the operation rate of specific end-of-pipe control technology, and η denotes the removal efficiency of specific end-of-pipe control technology for pollutants.

The formula for calculating CO_2 emissions parallels that used for atmospheric pollutant emissions. Nonetheless, given the limited widespread adoption of end-of-pipe control technologies, such as carbon capture, at the present stage, this parameter is not factored into the current calculations. The formula is articulated as follows:

$$E_i = \sum_j \sum_k \sum_m A_{i,j,k} X_{i,j,k,m} C_{i,j,k,m} \tag{2.2}$$

The parameters in the provided equation retain the same meanings as those in Eq. (2.1) for calculating atmospheric pollutant emissions.

From the aforementioned formulas, it becomes apparent that the foundational step in establishing urban-scale greenhouse gas and atmospheric pollutant emission inventories is to elucidate the composition of urban emission sources. Subsequently, relevant data, such as activity levels, emission factors, technological distribution, and information pertaining to end-of-pipe control technologies, should be systematically collected through surveys to facilitate the calculation of emissions.

2.1.3 Hierarchical Classification Methods of Emission Sources

The accurate identification of emission sources constitutes the primary step in the compilation of an emission inventory. It serves as the foundational basis for determining emission calculation methodologies and for the collection of pertinent data such as activity levels and emission factors. In this context, referencing the "Technical Manual for Compilation of Urban Atmospheric Pollutant Emission Inventory" [2], this book categorizes China's anthropogenic atmospheric pollutant sources into ten major classes: stationary fossil fuel combustion sources, process sources, mobile sources, solvent use sources, agricultural sources, dust sources, biomass combustion sources, storage and transportation sources, waste treatment sources, and other emission sources.

Recognizing variations in pollutant generation mechanisms and emission characteristics, each category of emission sources is further subdivided into four levels based on sectors/industries, fuels/products, combustion/process technologies, and end-of-pipe control technologies. This establishes a hierarchical classification system progressing from the first to the fourth level. The third-level emission sources primarily focus on identifying key emission sources characterized by high emissions and those significantly influenced by combustion/process technologies. For fuels and products with emissions less affected by combustion/process technologies, the third-level emission sources are not further subdivided; instead, the fourth-level emission source classification is established directly under the second-level emission sources.

Illustrating this with the example of stationary fossil fuel combustion sources, the four-level emission source classification structure is depicted in Fig. 2.3. The fourth level exclusively delineates dust removal technology.

For energy-intensive and environmentally impactful industrial sectors, such as thermal power, steel, cement, and glass, this study advocates the development of emission characterization models based on equipment and processes. In contrast, for

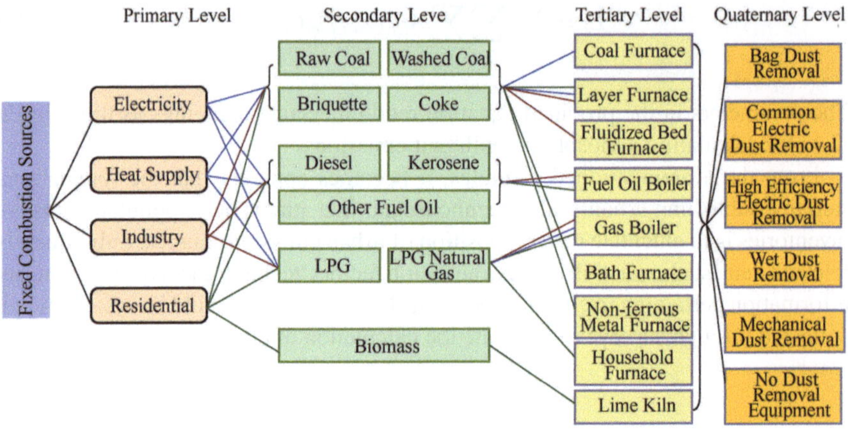

Fig. 2.3 Schematic diagram of the four-level emission source classification structure for stationary combustion sources

other industrial sectors, where emissions differ significantly among stationary fossil fuel combustion sources, process sources, and solvent use sources, distinct emission characterization methods are formulated based on equipment and technologies. Using the cement industry as an illustrative case, Fig. 2.4 portrays the dynamic emission characterization model rooted in the "product-production process-control technology" paradigm outlined in this publication. China's prevailing cement production technologies encompass shaft kilns, rotary kilns, and new dry-process kilns, each associated with distinct emission profiles. The principal pollutants emitted by the cement industry include particulate matter and NO_x, prompting the implementation of two categories of pollution control technologies: dust removal measures and denitrification measures. Dust removal measures encompass cyclone dust removal, wet dust removal, electrostatic dust removal, and baghouse dust removal. Denitrification measures include low nitrogen burners, selective non-catalytic reduction (SNCR), and selective catalytic reduction (SCR). Notably, these diverse control technologies exhibit variable removal efficiencies [3].

The initiation of the compilation process for urban atmospheric pollutant and greenhouse gas emission inventories necessitates a thorough survey of emission sources within the designated research area. Drawing from the distinctive characteristics of local industries and fuels/products, the selection of suitable first-level and second-level emission source categories is imperative, as per the hierarchical classification system of emission sources elucidated in this publication. Subsequently, it becomes crucial to ascertain the composition of local emission sources and identify subjects for the investigation and collection of activity level data. The data investigation and collection phase should encompass the combustion/process technologies associated with third-level emission source categories and the pollution end-of-pipe control technologies related to the fourth-level categories. During the data organization process, the determination of third-level and fourth-level emission source

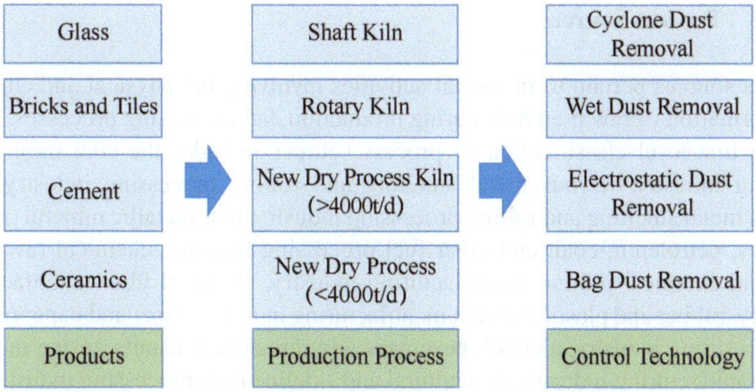

Fig. 2.4 Dynamic emission characterization model: illustrating the cement industry

categories covered by the emission inventory is contingent upon the characteristics of local emission sources. Leveraging the hierarchical classification system of local emission sources and the specifics of the data investigation, suitable methods and procedures for emission inventory compilation are established, grounded in the first-level emission source categories. Additionally, approaches and sources for acquiring calculation parameters are identified, taking into account the second- to fourth-level emission source categories.

With a focus on the ten categories of anthropogenic emission sources, the subsequent elucidation provides in-depth explanations of the four-level hierarchical classification system of emission sources as outlined in this book.

2.1.3.1 Stationary Fossil Fuel Combustion Sources

Stationary fossil fuel combustion sources encompass combustion equipment utilizing the heat generated from burning fossil fuels to provide energy for electricity generation, industrial production, and daily life needs.

The first-level classification of stationary fossil fuel combustion sources encompasses electricity generation, electricity supply, industrial heat production and supply, civil heat production and supply, gas production and supply, industrial boilers in the mining and manufacturing industries, urban civil sources, and rural civil sources. The second-level classification includes various fossil fuels such as coal, as well as different gaseous and liquid fuels. The third-level classification involves combustion equipment, such as boilers and stoves. The fourth-level classification comprises three types of pollution control measures: dust removal, desulfurization, and denitrification, in addition to scenarios where no control measures are implemented.

2.1.3.2 Process Sources

Process sources pertain to industrial activities involving the physical and chemical transformation of raw materials during production and processing processes.

The first-level classification of process sources includes the coal mining and washing industry, ferrous metal smelting and rolling processing industry, non-ferrous metal smelting and rolling processing industry, non-metallic mineral product industry, petroleum, coal, and other fuel processing industry, chemical raw materials and chemical product manufacturing industry, chemical fiber manufacturing industry, rubber and plastic product manufacturing industry, paper and paper product manufacturing industry, alcohol, beverages and refined tea manufacturing industry, food manufacturing industry, agricultural and sideline food processing industry, and textile industry. The second-level classification encompasses the main products or raw materials of these industries. The third-level classification involves major production processes and energy types. The fourth-level classification includes four categories of pollution control measures: dust removal, desulfurization, denitrification, and VOCs control technologies, along with scenarios where no control measures are implemented.

2.1.3.3 Mobile Sources

Mobile sources encompass diverse engine-driven passenger and freight transportation facilities and mechanical equipment capable of mobility.

The first-level classification of mobile sources comprises road mobile sources, encompassing passenger cars, freight trucks, and motorcycles, and non-road mobile sources, including construction machinery, agricultural machinery, small utility machinery, diesel generators, ships, railway locomotives, and civil aviation aircraft. The second-level classification involves primary vehicle types and fuel types for road mobile sources, as well as the main types of machinery for non-road mobile sources. The third-level classification encompasses emission levels under pre-China I and China I-VI vehicle emission standards. The fourth level is uniformly addressed in the absence of control measures.

2.1.3.4 Solvent Use Sources

Solvent use sources refer to both the industrial production and domestic sectors engaged in the production and application of organic solvents.

The first-level classification of solvent use sources includes artificial board manufacturing, printing and dyeing, various surface coatings, other industrial solvent uses, asphalt paving, pesticide use, and household solvent use. The second-level classification involves industrial solvent usage processes, such as printing, dyeing, architectural coatings, automotive painting, and surface coating, alongside civilian solvent usage processes, including insecticides, herbicides, and fungicides. The third-level classification encompasses various types of solvents, such as inks and paints. The fourth-level classification comprises VOCs control technologies and cases where no control measures are implemented.

2.1.3.5 Agricultural Sources

Agricultural sources encompass diverse activities within the agricultural production process that result in the emission of atmospheric pollutants.

The first-level classification of agricultural sources includes nitrogen fertilizer application, livestock and poultry farming, nitrogen-fixing plants, straw composting, soil background, and human feces. The second-level classification encompasses various types of nitrogen fertilizers, livestock and poultry, nitrogen-fixing plants, arable land, and rural populations. The third-level classification for livestock and poultry farming includes free-range, intensive farming, and grazing, while for other agricultural sources, the fourth-level classification is directly established based on the second level. The fourth-level classification is addressed in scenarios where no control measures are implemented.

2.1.3.6 Dust Sources

Dust sources are defined as emission sources releasing loose particulate matter from the ground surface in an unorganized and unregulated manner, propelled by natural forces or human activities, without passing through exhaust pipes.

The first-level classification of dust sources includes soil dust, road dust, construction dust, and stockpile dust. The second-level classification encompasses various land use types, road use types, construction types, and types of materials stockpiled. The third-level classification includes various soil textures, road paving conditions, construction phases, and material stacking procedures. The fourth-level classification involves urban dust control measures, such as watering, sweeping, and dust suppressant spraying.

2.1.3.7 Biomass Combustion Sources

Biomass combustion sources encompass the incineration processes of unmodified biomass materials utilized in boilers and stoves, in addition to occurrences such as forest fires, grassland fires, and the open burning of straw.

The first-level classification of biomass combustion sources consists of biomass fuels and open biomass burning. The second-level classification encompasses diverse biomass types, including straw, firewood, biomass pellets, livestock manure, grasslands, and forests. The third-level classification includes various combustion methods, such as biomass boilers and various stoves. Within the fourth-level classification for biomass boilers, there are three categories of pollution control measures:

dust removal, desulfurization, and denitrification, along with instances where no control measures are applied. Conversely, other biomass combustion sources are considered cases without control measures.

2.1.3.8 Storage and Transportation Sources

Storage and transportation sources pertain to the activities encompassing the gathering, storage, transportation, and marketing of volatile oil and gas products.

The first-level classification of storage and transportation sources covers oil and gas storage and transportation. The second-level classification includes the processes of storage, transportation, and sales of oil and gas products, such as crude oil, gasoline, diesel, and natural gas. The third-level classification remains unsegmented. The fourth-level classification encompasses the first, second, and third-stage oil and gas recovery processes at gas stations, along with instances where no oil and gas recovery is implemented.

2.1.3.9 Waste Treatment Sources

Waste treatment sources encompass locations where pollutants originate from the processing of industrial and domestic wastewater, as well as solid waste within centralized treatment and disposal facilities. Additionally, these sources include emissions of pollutants produced subsequent to flue gas denitrification.

The first-level classification of waste treatment sources encompasses wastewater treatment, solid waste treatment, and flue gas denitrification. The second-level classification includes wastewater, solid waste, and the flue gas resulting from denitrification. The third-level classification comprises solid waste composting and solid waste incineration, as well as SNCR and SCR in flue gas denitrification. The fourth-level classification is addressed in instances where no control measures are implemented.

2.1.3.10 Other Emission Sources

Other emission sources encompass the compilation of atmospheric pollutant emission sources not accounted for in the aforementioned classifications.

The first-level classification of other emission sources presently includes solely restaurant cooking fumes. The second-level classification encompasses kitchen cooking fumes. The third-level classification is no longer further subdivided. The fourth-level classification includes fume purifiers and instances where no control measures are applied.

2.1.4 Methods of Emission Inventory Compilation

The fourth-level emission source classification stands as the foundational calculation unit for emission inventory compilation. Emission calculation methods fall into two categories: the online monitoring method and the pollution production and discharge coefficient method. The treatment of point sources and area sources varies based on data availability. Point sources are emission sources with available precise spatial locations and activity levels. In calculating emissions from point sources, latitude and longitude coordinates, along with activity levels, must be obtained for each individual pollution-emitting device. Area sources represent an amalgamation of emission sources where it is challenging to precisely delineate emission locations and activity levels. In emission inventories, area sources are generally presented as the overall emissions for cities or districts. When calculating emissions from area sources, it is crucial to designate the smallest administrative unit participating in the calculation (typically districts or streets) as the basis for obtaining activity levels. In the process of compiling the emission inventory, it is imperative to articulate the methodology for calculating each fourth-level emission source classification. Industrial enterprises should, to the greatest extent possible, compute emissions from each individual pollutant discharge equipment based on the point source. For emission sources with limited statistical data, such as civilian and agricultural sources, emissions can be computed at the level of the smallest administrative unit based on the area source.

The subsequent sections delineate the methodologies employed in compiling emission inventories for select key sectors.

2.1.4.1 Thermal Power Sector

The principal apparatus responsible for production and pollutant discharge within the thermal power sector comprises thermal power units, where pollutants are generated during the combustion process of fossil fuels. In accordance with the previously introduced emission source classification method, the first-level classification for the thermal power sector is denoted as stationary fossil fuel combustion sources. The second-level classification is further differentiated into electricity production and industrial/civil heat production and supply, based on the fuels employed for electricity generation and heat supply. The third-level classification encompasses various fossil fuels, such as coal, petroleum, and natural gas. Moving to the fourth level, an array of power generation equipment is considered, including coal-fired boilers, fluidized bed boilers, automatic grate layer furnaces, oil-fired boilers, and gas-fired boilers. A comprehensive emission inventory for the thermal power sector is systematically established from the first-level classification to the fourth-level classification. Notably, emissions from electricity generation and heat supply are calculated distinctly, as articulated below:

$$E = A \times \mathrm{EF} \times (1 - \eta) \tag{2.3}$$

where A signifies the fuel consumption of each specific pollutant discharge apparatus, EF represents the pollutant generation factor, and η denotes the removal efficiency of pollutant control measures. Notably, in thermal power plants equipped with carbon capture and storage systems, this parameter assumes a value for CO_2, whereas in alternative scenarios, it conventionally defaults to 0 for CO_2.

The thermal power sector gathers activity level data on a per-unit basis, including parameters such as unit installation capacity, operational hours of power units, operational hours of desulfurization facilities, and concentrations from online monitoring. Coal sulfur and ash content are determined using batched coal quality data, with the average content calculated through a weighted method. In the case of combined heat and power enterprises, fuel consumption for electricity generation and heat supply should be separately obtained from power plant production reports to compute emissions for electricity generation and heat supply. The collection of activity level data should be integrated into the environmental data statistical system, drawing relevant information from databases related to total verification, environmental statistics, and pollution source censuses. On-site surveys serve as supplementary sources. Details on unit installed capacity, commissioning date, fuel consumption, sulfur and ash content, boiler type, combustion method, types of desulfurization and denitrification facilities, and overall removal efficiency can be directly obtained from comprehensive verification data. Data on operating hours of pollutant discharge equipment and types of dust removal facilities can be sourced from pollution source censuses and environmental statistics. Building on this foundation, electricity and heat production and supply departments should conduct additional on-site investigations to enhance activity level data. Particular attention should be paid to data on the operating hours of pollution control facilities, online monitoring concentration data, enterprise production reports, and batched coal quality data.

Emission factors are preferably derived through direct measurements. Alternatively, online monitoring data from pollution sources, encompassing fuel composition, pollutant concentrations in flue gas, and oxygen content, can be utilized for calculations. In regions where conducting observations or collecting monitoring data from emission sources is challenging, emission factors for the fourth-level emission source classification can be determined through the material balance method and by referencing emission factor databases.

The overall desulfurization (denitrification) efficiency in the thermal power sector is calculated by multiplying the desulfurization (denitrification) facility operation rate by the desulfurization (denitrification) efficiency. Specifically, the desulfurization (denitrification) facility operation rate represents the proportion of the operating time of the desulfurization (denitrification) facility to the elapsed time of the power generation unit. The desulfurization (denitrification) efficiency can be obtained by referencing emission factor databases provided in the appendix of the "Technical Manual for Compilation of Urban Atmospheric Pollutant Emission Inventory". Emission factors for other pollutants and removal efficiencies of pollution control facilities can similarly be derived by referencing the same databases based on the fourth-level classification to which the emission source belongs.

2.1.4.2 Industrial Sector

The industrial sector encompasses various emission sources, notably stationary fossil fuel combustion sources, process sources, and solvent use sources. The subsequent discussion primarily centers on the first two categories of emission sources.

1. Stationary fossil fuel combustion sources

   Illustrated with examples from mining and manufacturing industries, the emission calculation formula for the fourth-level emission source classification is articulated as follows:

$$E = A \times \mathrm{EF} \times (1 - \eta) \tag{2.4}$$

Where A represents the fuel consumption related to the fourth-level emission source classification. In the context of point sources, A signifies the activity level of the pollutant discharge equipment. Conversely, for area sources, A represents the activity level of the smallest administrative unit considered in the emission inventory. EF denotes the pollutant or CO_2 generation factor, while η symbolizes the removal efficiency of pollutant control measures, typically set to 0 in the case of CO_2.

The survey of activity levels encompasses diverse industries within the mining and manufacturing sectors, with distinct considerations for point sources and area sources. Point sources involve industrial boilers with a capacity of 35 steamed tons per hour or above, whereas those with a capacity of 10 steamed tons per hour or below are classified as area sources. Industrial boilers with a capacity of 10–35 steamed tons per hour can be categorized as point or area sources based on local circumstances.

Individual investigations should be conducted for point sources to gather information on the activity levels of pollutant discharge equipment. The data to be collected include the latitude and longitude of pollutant discharge equipment, fuel type, boiler type, fuel consumption, pollution control facility type, and temporal variation curve of production load. For coal-fired boilers, the ash content and sulfur content of the coal are required. The fourth-level emission source classification for each pollutant discharge facility should be determined based on its fuel type, boiler type, and pollution control facility type. Facilities equipped with continuous monitoring systems for flue gas emissions necessitate obtaining hourly average pollutant emission concentrations, hourly average flue gas emissions, and total production hours for each monitoring section of the flue.

In the case of area sources, activity levels should be collected according to the smallest administrative units in the emission inventory. Initially, obtaining the activity levels for the second-level emission source classification, including administrative district names, zoning codes, and fuel consumption by sub-sector is crucial. Subsequently, through on-site surveys and analog surveys, the technical distribution ratios for the third-level and fourth-level emission source classifications are derived, facilitating the calculation of activity levels for the fourth-level emission source classification.

The investigation and collection of activity level data should be seamlessly integrated into the environmental data statistical system, leveraging relevant enterprise information from databases for environmental statistics and pollution source census, supplemented and enriched through on-site surveys. For point sources' activity levels, essential data such as fuel consumption, coal ash content and sulfur content, boiler type, fuel type, and dust removal facility type can be directly derived from the pollution source census database. However, data like the latitude and longitude of pollutant discharge facilities and other pollution control facility types require supplementation and enhancement based on on-site surveys. Regarding area sources' activity levels, sub-sector fuel consumption for each district and county, as well as the proportions of various types of boilers and dust removal facilities, can be extracted from the pollution source census database. In the case of area sources, targeted surveys of activity levels for small-and medium-sized boilers should be conducted within the emission inventory preparation area to address missing and omitted data in the environmental data statistical system.

Preferably, emission factors are obtained through direct measurements. Alternatively, calculations can be based on online monitoring data from pollution sources, encompassing parameters such as fuel composition, pollutant concentrations in flue gas, and oxygen content. In regions facing challenges in conducting observations or obtaining monitoring data from emission sources, derivation of emission factors for the fourth-level emission source classification can be achieved through the material balance method and by consulting emission factor databases outlined in the appendix of the "Technical Manual for Compilation of Urban Atmospheric Pollutant Emission Inventory".

2. Process sources

   Taking the cement industry as an example, the formula for computing pollutant emissions in the fourth-level emission source classification is articulated as follows:

$$E = A \times \mathrm{EF} \times (1 - \eta) \tag{2.5}$$

where A represents the activity level corresponding to the fourth-level classification of emission sources, such as clinker or cement production. In the case of point sources, A signifies the activity level specific to each production line, while for area sources, A denotes the activity level at the smallest administrative unit within the emission inventory. EF denotes the generation factor for various pollutants and CO_2. In this context, CO_2 is generated through the high-temperature decomposition of raw materials (e.g., $CaCO_3 \rightarrow CaO + CO_2$) in the cement production process, excluding carbon emissions already considered in stationary fossil fuel combustion sources. Symbol η represents the removal efficiency of pollutant control measures, typically set at 0 for CO_2.

The survey scope for activity level data within the cement industry encompasses all industrial enterprises involved in clinker and cement production within the designated area for emission inventory preparation. Emission sources within this context should be considered as point sources.

Specifically, for the cement industry, a detailed investigation of emission sources is imperative to gather activity level information from facilities responsible for pollutant discharge. The requisite data includes the geographical coordinates (latitude and longitude) of pollutant discharge equipment, production facility type, production scale, clinker and cement production quantities, details of pollution control measures implemented at each emission point, pollutant removal efficiency, and the temporal variation curve of production load. The fourth-level classification of emission sources can be determined by considering the process type and pollution control facility type associated with each pollutant discharge facility. For facilities equipped with continuous monitoring systems for flue gas emissions, it is essential to acquire information on the hourly average pollutant emission concentrations, hourly average flue gas emissions, and total production hours for each monitoring section of the flue.

The survey and compilation of activity level data necessitate seamless coordination with comprehensive cement enterprise information within the total verification and accounting system. Concurrently, this effort should be harmonized with pertinent data derived from pollution source censuses and environmental statistics databases, subsequently augmented and refined through meticulous on-site investigations. Concerning activity level data for point sources, crucial information such as the clinker production facility type, production scale, product output, denitrification facility type, and denitrification efficiency can be directly extrapolated from the total verification and accounting system. The production scale and product output associated with cement grinding can be ascertained through a thorough amalgamation of data from the total verification and accounting system, pollution source census database, and environmental statistics database. Some categories of dust removal facilities within cement enterprises can be deduced from the pollution source census database. Details such as the geographical coordinates (latitude and longitude) of pollutant discharge equipment and other pertinent information not readily available should be supplemented and refined through on-site surveys.

It is highly preferable to gather activity level data that closely aligns with the designated target year of the emission inventory. In instances where activity level data for the specified target year is unavailable, recourse to data from proximate years is acceptable, with requisite adjustments based on prevailing socioeconomic development conditions.

The acquisition of pollutant generation factors is optimally achieved through direct measurements. Alternatively, these factors can be computed by collating online monitoring data from pollution sources, encompassing variables such as fuel composition, pollutant concentrations in flue gas, and oxygen content. In regions where the execution of observations or the collection of monitoring data from emission sources poses challenges, emission factors for the fourth-level emission source classification can be derived using the material balance method. Reference to the emission factor databases provided in the appendix of the "Technical Manual for Compilation of

Urban Atmospheric Pollutant Emission Inventory" is recommended as well. CO_2 emission factors can be sourced from the "Provincial Greenhouse Gas Inventory Compilation Guidelines", with the option of direct measurements in regions where such procedures are feasible.

2.1.4.3 Transportation Sector

In this context, road mobile sources serve as an illustrative example to elucidate the methodology employed in compiling the emission inventory for the transportation sector. Emissions are computed according to the fourth-level emission source classification and are designated as area sources. The expression of these emissions is delineated as follows:

$$E = P \times \mathrm{EF} \times \mathrm{VKT} \tag{2.6}$$

where P represents the count of motor vehicles, EF signifies emission factors derived from mileage traveled, and VKT denotes the annual vehicle kilometers traveled.

The collection of activity level data for road mobile sources encompasses diverse motor vehicle types, their geographical distribution, vehicle quantities, registration years, and emission control levels. Information concerning vehicle types, quantities, and registration years can be sourced from the traffic administration department. Details pertaining to vehicles at different emission control stages can be obtained from the database of vehicle environmental protection inspection management systems within the relevant ecological and environmental protection departments. In instances where vehicle environmental protection inspection data is unavailable, reference to the local implementation progress of motor vehicle emission standards can facilitate the identification of emission control stages based on vehicle registration years. The fourth-level emission source classification for road mobile sources can be established by considering fuel type, vehicle type, and emission control level. The annual vehicle kilometers traveled (AVKT) necessitate on-site surveys, specifically derived from local annual motor vehicle inspection data, where the cumulative mileage and vehicle usage years are tallied by vehicle type to calculate the AVKT. In circumstances where annual motor vehicle inspection data is absent, on-site investigations can be conducted to determine vehicle type-specific AVKT through random sampling visits to prominent parking facilities. In areas where on-site surveys have not been conducted, the national AVKT figures provided in the appendix of the "Technical Manual for Compilation of Urban Atmospheric Pollutant Emission Inventory" can be consulted.

The emission factor for motor vehicle tailpipe emissions undergoes modification by amalgamating the baseline emission factor with actual conditions, as depicted in the subsequent formula:

$$\mathrm{EF} = \mathrm{BEF} \times \phi \times \gamma \times \lambda \times \theta \tag{2.7}$$

where BEF designates the baseline emission factor, φ signifies the environmental correction factor, γ denotes the average speed correction factor, λ represents the deterioration correction factor, and θ denotes the correction factor accounting for other usage conditions (including load factor, fuel quality, etc.).

The baseline emission factor (BEF) serves as a representation of vehicle emission levels under typical driving conditions, average fuel quality, and ambient environmental circumstances. The conditions associated with the baseline emission factors for diverse motor vehicle types are delineated as follows: typical urban conditions (30 km/h); national average deterioration conditions in 2014; a temperature of 15 °C, relative humidity of 50%, and low altitudes; gasoline without ethanol blending and with a sulfur content of 50 ppm, diesel with a sulfur content of 350 ppm; and a load factor of 50% for diesel vehicles. The baseline emission factors for various motor vehicle types in road mobile sources are available in the appendix of the "Technical Manual for Compilation of Urban Atmospheric Pollutant Emission Inventory".

The environmental correction factor (φ) is introduced to account for the influence of environmental factors (temperature, humidity, altitude, etc.) on vehicle emissions. It is computed through the temperature correction factor, humidity correction factor, and altitude correction factor, expressed as follows:

$$\phi = \phi_{\text{Temp}} \times \phi_{\text{RH}} \times \phi_{\text{Height}} \tag{2.8}$$

In this context, φ_{Temp} denotes the temperature correction factor, φ_{RH} represents the humidity correction factor, and φ_{Height} signifies the altitude correction factor. The "Technical Manual for Compilation of Urban Atmospheric Pollutant Emission Inventory" appendix furnishes temperature correction factors, humidity correction factors, and altitude correction factors, which can be chosen based on the specific local conditions. Moreover, for motor vehicle types and pollutants that exhibit minimal sensitivity to environmental factors and are not explicitly enumerated in the appendix, an environmental correction factor of 1 may be applied.

The introduction of the average speed correction factor (γ) aims to capture the impact of driving speed on vehicle emissions. The "Technical Manual for Compilation of Urban Atmospheric Pollutant Emission Inventory" furnishes γ values for five speed intervals: <20 km/h, 20–30 km/h, 30–40 km/h, 40–80 km/h, and > 80 km/h. Selection of the appropriate γ values should be based on the prevailing local road conditions.

The deterioration correction factor (λ) is incorporated to accommodate the declining condition of vehicles and the consequent rise in emission factors with increasing mileage. The "Technical Manual for Compilation of Urban Atmospheric Pollutant Emission Inventory" provides λ values for various motor vehicle types for the years spanning from 2015 to 2018, with 2014 serving as the base year. λ values for diverse motor vehicle types prior to 2014 are standardized to 1. During the compilation of the emission inventory, the λ values should be chosen in accordance with the target year.

The correction factor for other usage conditions (θ) is introduced to account for the influence of sulfur content in fuel, ethanol blending, and the load of diesel vehicles on vehicle emissions. The determination of θ values can be made by consulting the appendix of the "Technical Manual for Compilation of Urban Atmospheric Pollutant Emission Inventory", taking into consideration the specific local conditions.

In summary, the collection of necessary data involves factors such as city temperature, humidity, altitude, road average vehicle speed, sulfur content in fuel, ethanol blending, and the load of diesel vehicles. Information on temperature, humidity, and altitude can be sourced from local environmental statistical yearbooks. The traffic management department can provide data on road average vehicle speed. Fuel quality information can be obtained through surveys of the relevant business departments responsible for fuel sales. Diesel vehicle load data can be acquired through surveys conducted at toll collection stations.

The process of compiling the emission inventory necessitates the calculation of diverse correction factors by consulting the appendix of the "Technical Manual for Compilation of Urban Atmospheric Pollutant Emission Inventory" in accordance with the prevailing actual circumstances. In instances where the actual conditions deviate from the parameters outlined in the appendix, the corresponding correction factors for those conditions can be determined through linear interpolation.

2.1.4.4 Civil Sector

For the fourth-level emission source classification of civil sources, the formula for calculating emissions is as follows:

$$E = A \times \mathrm{EF} \times (1 - \eta) \tag{2.9}$$

where A represents the fuel consumption for the fourth-level emission source classification, EF denotes the pollutant generation factor, and η represents the removal efficiency of pollutant control facilities.

Civil sources are typically regarded as area sources. The investigation and collection of activity level data for the smallest administrative units in the emission inventory are imperative. Energy consumption data for various types of energy sources in the civil sector can be sourced from the local statistical department. If such data is unavailable locally, it can be obtained by deducting electricity and industrial energy consumption from the total local energy consumption. Alternatively, activity level data for civil sectors at a higher administrative level can be employed, with subsequent weighting and allocation to the emission inventory compilation administrative units based on parameters such as population density. The acquired activity level data typically falls within the second-level emission source classification. Through various methodologies, including on-site sampling surveys and analog surveys, technical proportions for the third-level and fourth-level emission source classifications, such as the proportion of decentralized heating boilers and domestic stoves, can be determined. This aids in establishing activity levels for the fourth-level emission source classification.

To acquire activity level data from civil sources, equal emphasis should be placed on statistical surveys and on-site investigations. In the absence of detailed macro-level statistical data, deficient activity level data can be supplemented through on-site investigations. Information about civil pressured boilers can be obtained from local market supervision and management departments, while data for atmospheric pressure boilers requires on-site surveys in decentralized heating areas. On-site investigations for activity level data of civil sources should particularly focus on civil bulk coal, with corresponding fuel consumption data obtainable through surveys conducted by statistical departments, development and reform commissions, and agricultural departments. However, for rural areas with limited statistical data, on-site investigations are indispensable.

Emission factors for civil boilers can be derived by referencing the treatment of emission factors for industrial boilers in the mining and manufacturing industries. Cities with well-established foundations are encouraged to conduct actual emission factor tests for local typical civil fuels and stove types. In the absence of such tests, cities can opt for emission factors provided in the appendix of the "Technical Manual for Compilation of Urban Atmospheric Pollutant Emission Inventory". Each city should select emission factors that align with the local emission source hierarchical classification system based on the investigation of activity level data.

2.1.5 Methods for Data Acquisition

The essential data for compiling an emission inventory can be broadly categorized into two main types: activity level data and emission factors. Both not only serve as the foundational elements for calculating emissions but also constitute the primary origins of uncertainty in emission results.

2.1.5.1 Activity Level Data

In the process of collecting activity level data, it is imperative to develop individual survey plans tailored to each fourth-level emission source classification. Moreover, the creation of survey checklists, establishment of survey procedures, and specification of approaches for acquiring activity level data are essential components of this undertaking. The survey collection process for activity level data can be seamlessly integrated with existing data statistical systems, drawing relevant information from databases dedicated to environmental statistics, pollution source census, and pollution discharge declarations.

The survey of activity level data should diligently aim to gather data corresponding to the base year to the fullest extent possible. In scenarios where activity level data for the base year is unavailable, recourse to data from adjacent years is permissible, with adjustments made based on prevailing socioeconomic development conditions. In instances where a city possesses comprehensive energy balance sheets and industrial product output data, these datasets should function as overarching constraints for the activity levels of urban emission sources in the corresponding year. For sources exhibiting discrepancies, a thorough analysis and verification process should be initiated, followed by necessary adjustments to enhance data accuracy and reliability.

2.1.5.2 Emission Factors

Through a fusion of literature research and emission measurements, this publication constructs an emission factor database tailored to major atmospheric pollutants and greenhouse gases in China. The utilization of this database is integral to the compilation of urban-scale emission inventories. The subsequent overview provides a concise outline of the methods employed for obtaining emission factors.

1. Atmospheric pollutants

   The methodologies for acquiring emission factors for atmospheric pollutants generally involve three primary approaches: direct measurements, the material balance method, and literature research. Cities can select the most suitable method based on their specific operational context.

   In regions with favorable conditions, conducting actual emission factor tests for local key emission sources is recommended. These tests yield emission factors that accurately capture the characteristics of local emission sources. It is crucial that these tests be executed under the normal operating conditions of the emission source to obtain emission levels representative of typical scenarios.

   The material balance method involves deriving generation factors through a meticulous analysis of input and output materials. This analysis is coupled with the removal efficiency of pollution control equipment or measures, culminating in the determination of final emission factors. The material balance method is particularly applicable for estimating emission factors for SO_2 and particulate matter emanating from large- and medium-sized coal-fired equipment, as well as for VOCs originating from solvent use sources.

   Given constraints in testing conditions and method applicability, emission factors for certain sources need to be acquired through literature research. The literature encompasses not only published research papers and academic reports but also established emission factor databases. Noteworthy international databases include the Compilation of Air Pollutant Emission Factors (AP-42) published by the U.S. and the EMEP/EEA air pollutant emission inventory released by the EU. On the domestic front, emission factor databases can be sourced from the "Technical Manual for Compilation of Urban Atmospheric Pollutant Emission Inventory", the "Manual of Emission Source Statistical Investigation, Pollutant Generation and Discharge Accounting Methods and Coefficients", or other pertinent literature and documents. Guided by the emission source hierarchical classification system delineated in this book, emission inventory compilers can systematically conduct literature research for each emission source, tailoring their approach to specific circumstances. This involves collecting both domestic and international emission factor test results and databases. Additionally, establishing preferred criteria for selecting emission factors and conducting a comprehensive evaluation of the quality grade of emission factor test values are imperative steps to ultimately determine the applicable emission factors. In instances where determining emission factors proves challenging, the "Atmospheric Pollutant Emission Inventory Compilation and Analysis System" offers a viable alternative. This includes utilizing built-in generation factors and removal efficiency data derived from pollution control measures as a substitute approach.

2. Greenhouse gases

The methods for obtaining greenhouse gas emission factors primarily involve research tests, enterprise production tests, and reliance on international or domestic recommended data.

Historically, previous studies frequently depended on emission factors published in the "IPCC Guidelines for National Greenhouse Gas Inventories" for compiling greenhouse gas emission inventories. However, it is noteworthy that the emission factors provided by the IPCC may not accurately reflect the actual emissions from specific sources in China.

In a study conducted by Liu et al., the focus was on the contemporary energy structure and CO_2 emission characteristics in China. The researchers systematically collected over 20,000 sets of diverse energy samples, encompassing sample categories that accounted for more than 95% of China's fossil fuel consumption. Through a meticulous analysis of these energy samples, the research team computed the emission factors for various energy types in China. The findings of the study revealed that China's coal emission factor is 45% lower than the recommendation provided by the IPCC. In contrast, the emission factors for oil and gas closely align with the IPCC recommendations. Notably, China's coal exhibited a higher ash content, averaging 27%, compared to the U.S. average of 14%. This disparity indicates that the carbon content in China's coal is considerably lower relative to developed countries and the global average. The study's measurements indicated an average carbon content of 54% in China's coal, whereas the default value suggested by the IPCC is 75%. Generally, a lower carbon content corresponds to a lower caloric value. The study determined the calorific value of China's coal to be 20.6 PJ/Mt., in contrast to the IPCC recommendation of 28.2 PJ/Mt., as illustrated in Fig. 2.5 [4].

The carbon oxidation rate denotes the proportion of carbon in the fuel oxidized into CO_2 during the combustion process. In a comprehensive investigation into fuel combustion in China, this study specifically focused on the carbon oxidation rate and made notable findings. The average carbon oxidation rate for coal in various industrial technology types in China was determined to be 92%, which is lower than the IPCC recommendation of 98%. Calculated oxidation coefficients for oil and natural gas exhibited no significant deviation from the IPCC recommendation. The study encompassed a thorough survey of 135 major combustion technologies, ultimately obtaining a weighted average of the carbon oxidation rate for coal combustion in China. The analysis revealed a significant deviation from the IPCC recommendation, primarily attributable to the fact that in China, only 40% of coal is consumed in thermal power generation. The majority of coal is utilized in various industrial boilers, reflecting the diverse range of industrial technology levels in China. Substantial differences in combustion technologies and boiler scales were observed across different industries and regions. In contrast, the IPCC bases its recommendations on the experience and technology levels of developed countries, where 90% of coal is used for combustion in thermal power generation. Thermal power generation typically boasts higher combustion efficiency and oxidation coefficients.

The cement industry contributes significantly to CO_2 emissions through the calcination process, where the decomposition of calcium carbonate and magnesium carbonate occurs. This process constitutes a substantial portion of the overall carbon emissions from industrial processes in China. To address this, extensive on-site measurements were conducted on cement production lines throughout China to acquire detailed carbon emission factors and related parameters for the cement production process. As a result of these measurements, the cement-clinker factor for China was calculated to be approximately 60%, a figure considerably lower than the international standard of 95% endorsed by various organizations. This discrepancy is attributed to the substantial cement production volume in China and the mixed use of various cement types in many plants, in contrast to developed countries that predominantly use silicate cement, commonly known as Portland cement. Consequently, the carbon emission factor for the cement production process in China is estimated to be approximately 30% lower than the value recommended by the IPCC.

Given the disparities between IPCC recommendations and the specific circumstances in China, this book advocates prioritizing the utilization of domestic emission factor databases. These include resources such as "The People's Republic of China National Greenhouse Gas Inventory", "2005: The People's Republic of China National Greenhouse Gas Inventory", "Compilation Guidelines for Provincial Greenhouse Gas Inventory", and "Guidelines for the Development of City's Greenhouse Gas Inventory in China". In instances where a city's production processes, technology levels, or fuel types significantly deviate from the national average, independent emission factor research is recommended. This research should be based on measured carbon content, lower calorific value, and carbon oxidation rate of fuels to ensure accuracy and relevance to local conditions.

2.1.6 Emission Space Allocation

Emission space allocation involves the establishment of emission space grids to fulfill the input requirements of air quality models. Ideally conducted at the fourth-level emission source classification, this process treats point sources and area sources

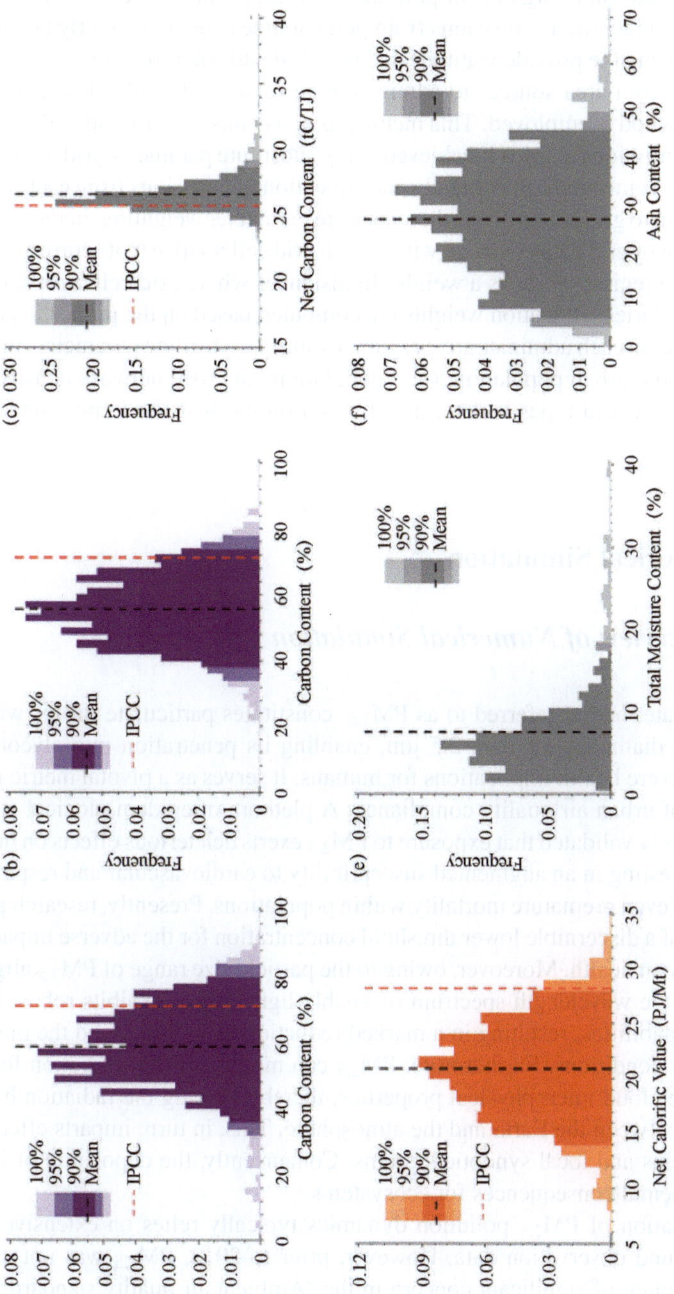

Fig. 2.5 Analysis of coal characteristics in China

differently. In the survey of activity level data for point sources, it is imperative to gather the latitude and longitude of pollution discharge equipment, accompanied by a clear location identifier. Emissions from point sources are then directly positioned on the grids using the provided latitude and longitude information. For the allocation of emissions from area sources to administrative regions, the substitute parameter weighting method is employed. This method assigns emissions to grid cells, and the mapping of emissions to grids is achieved using substitute parameter grid data highly correlated with emissions. Specifically, the allocation of emissions from each administrative region to grid cells utilizes the substitute parameter weighting method, where the proportion of grid data coverage within each grid cell relative to its corresponding administrative region serves as a weight. In instances where grid cells cross administrative boundaries, allocation weights are computed based on the proportion of the area they cover in each administrative region. Common substitute parameters include total population, urban population, rural population, and road network density. The selection of grid data types is determined based on the first-level emission source classification.

2.2 Numerical Simulation

2.2.1 *Overview of Numerical Simulation*

Fine particulate matter, referred to as $PM_{2.5}$, constitutes particulate matter with an aerodynamic diameter less than 2.5 μm, enabling its penetration into alveoli and instigating severe health implications for humans. It serves as a pivotal metric in the assessment of urban air quality compliance. A plethora of epidemiological studies has conclusively validated that exposure to $PM_{2.5}$ exerts deleterious effects on human health, manifesting in an augmented susceptibility to cardiovascular and respiratory diseases, and even premature mortality within populations. Presently, research posits the absence of a discernible lower threshold concentration for the adverse impacts of $PM_{2.5}$ on human health. Moreover, owing to the particle size range of $PM_{2.5}$ aligning closely with the wavelength spectrum of visible light, $PM_{2.5}$ exhibits robust light-scattering capabilities, resulting in a marked reduction in visibility and the onset of haze weather conditions. Furthermore, $PM_{2.5}$ can modify surface radiation forcing and influence cloud microphysical properties, thereby altering the radiation budget equilibrium between the Earth and the atmosphere. This, in turn, imparts effects on climate systems and local synoptic systems. Concurrently, the deposition of $PM_{2.5}$ yields detrimental consequences for ecosystems.

The evaluation of $PM_{2.5}$ pollution dynamics typically relies on extensive time series of ground observation data. However, prior to 2012, $PM_{2.5}$ was not classified as a pollutant of significant concern in the "Ambient air quality standards (GB 3095—1996)" and its subsequent amendments. Consequently, conventional ambient air quality monitoring did not encompass $PM_{2.5}$ concentrations in its purview, and governmental environmental quality reports omitted data on $PM_{2.5}$ concentrations. Only a select few research institutions conducted $PM_{2.5}$ monitoring [5]. However, these studies were hampered by limited spatial coverage, disparate timeframes, and inconsistent observation methodologies, rendering the quantitative assessment of air quality variations challenging.

Air quality models, specifically atmospheric chemical transport models, represent potent instruments for delineating the spatiotemporal distribution of ambient $PM_{2.5}$ concentrations. These models furnish continuous spatiotemporal information regarding near-surface $PM_{2.5}$ concentrations and find extensive utility in air quality assessments. The ambient $PM_{2.5}$ concentrations are contingent upon emission rates of atmospheric pollutants, meteorological conditions, and are subject to the influence of various atmospheric physical and chemical processes. Grounded in contemporary scientific understanding of these processes, air quality models employ numerical simulation methods to parameterize physical and chemical aspects related to air quality, facilitating a quantitative depiction of the transport and transformation of atmospheric pollutants. Despite computational intensity and limitations in accuracy, air quality models offer simulated results for comprehensive spatiotemporal series, an achievement often elusive with other datasets. Moreover, these models can be employed to design sensitivity experiments by adjusting emission inventories and meteorological parameters, contributing significantly to policy assessment research.

The exploration of air quality models commenced in the 1960s, and as scholarly investigations progressed, comprehension of atmospheric processes deepened gradually. These models have evolved into the third generation. The initial generation of air quality models drew from Gaussian dispersion models and operated on simplistic linear mechanisms. Notably, they lacked chemical reaction modules or relied on rudimentary ones. Representative models from this era include the Industrial Source Complex Model, California Puff Model, and Empirical Kinetics Modeling Approach. Subsequently, the second-generation air quality models, grounded in Eulerian grid models, notably enhanced the performance of chemical modules compared to their predecessors. While these models were instrumental in studying "single-pollutant issues", such as acid deposition and photochemical pollution, they lacked the coupling between atmospheric pollution processes. Representative models from the second generation include the Urban Airshed Model, Regional Oxidant Model, and Regional Acid Deposition Model. However, both the first and second generations were deficient in providing a detailed description of the formation mechanisms responsible for atmospheric $PM_{2.5}$, thus impeding an accurate simulation of its spatiotemporal distribution. In contrast, third-generation air quality models, founded on the concept of "one atmosphere", integrate the physical and chemical processes of multiple pollutants within a unified framework. This unification allows these models to be instrumental in studying "multi-scale multi-pollutant issues", significantly enhancing their reliability for policy support. Presently, most research employs third-generation air quality models for air quality simulation and policy assessment. Prominent models in use include the Comprehensive Air Quality Model with Extensions, Community Multi-Scale Air Quality Model (CMAQ),

Weather Research and Forecasting Model Coupled to Chemistry, and Goddard Earth Observing System-Chemistry.

2.2.2 WRF-CMAQ Model

To undertake an analysis of urban air quality compliance, it is imperative to employ an air quality model to conduct numerical simulations for $PM_{2.5}$ concentrations. These simulations, facilitated by iterative processes, enable the determination of atmospheric environmental capacity. The selected air quality model for this study must satisfy the following three criteria:

1. It should comprehensively account for the physical transport and chemical transformation processes among diverse pollutants, enabling the simulation of the synergistic effects of multiple pollutants.
2. The model should have the capability to simulate atmospheric environmental phenomena across local, regional, and national scales.
3. It must have the ability to concurrently simulate various atmospheric pollution processes, encompassing SO_2, NO_2, PM_{10}, $PM_{2.5}$, O_3, and acid rain, with a particular emphasis on compound regional atmospheric pollution processes.

The CMAQ model, founded on the principle of "one atmosphere", incorporates intricate physical and chemical processes, enabling the simultaneous simulation of complex atmospheric environments characterized by multiple scales, species, and processes. As the predominant numerical model aligning with the air quality simulation criteria of this study, the CMAQ model stands out. The utilization of the CMAQ model necessitates meteorological data for propulsion. Consequently, in this study, the Weather Research and Forecasting-Community Multi-Scale Air Quality (WRF-CMAQ) air quality simulation system is implemented to scrutinize urban air quality conditions under various combinations of measures.

The Weather Research and Forecasting (WRF) system, designed to cater to the demands of atmospheric science research and operational forecasting, stands as the next-generation mesoscale numerical weather prediction system under the guidance of the National Center for Atmospheric Research (NCAR). WRF encompasses two dynamic solving systems: the Advanced Research WRF (ARW), developed by NCAR, and the Non-Hydrostatic Mesoscale Model (NMM), developed by the National Centers for Environmental Prediction (NCEP). Presently, the preeminent choice for research applications is the WRF version employing the ARW dynamic solving system for the simulation of meteorological parameters. In the context of this study, the same version of WRF is employed to simulate variations in meteorological parameters in China, providing essential meteorological field inputs for the CMAQ model.

The CMAQ model represents a third-generation air quality model developed under the auspices of the U.S. Environmental Protection Agency (USEPA). Its modular structure includes components such as the boundary conditions processor

(BCON), initial conditions processor (ICON), photolysis rate processor (JPROC), meteorology-chemistry interface processor (MCIP), and chemical transport module (CCTM), as illustrated in Fig. 2.6. Within this framework, CCTM assumes a central role, functioning as the nucleus of the CMAQ model. It supplies input data and pertinent parameters for the dispersion and transport of pollutants in the atmosphere, meteorological-chemical processes, aerosol chemistry processes, liquid-phase chemistry processes, cloud chemistry processes, and dynamic processes. The ancillary modules primarily serve to furnish input data and relevant parameters for CCTM. CCTM possesses the capability to generate hourly concentrations of various gaseous pollutants and aerosol components, along with hourly visibility and dry/wet deposition outputs. The meteorological data requisite for the CMAQ model are simulated through the WRF model. Subsequently, the outputs from the WRF model undergo processing by MCIP, serving as meteorological field inputs for the CMAQ model.

The foundational design concept of the CMAQ model, illustrated in Fig. 2.7, revolves around the principle of "one atmosphere". This design tightly interlinks diverse pollutants and pollution issues within the atmosphere through chemical reactions. It underscores the holistic understanding of atmospheric physical and chemical processes, enabling the simultaneous simulation of multiple pollutants and pollution issues. These encompass photochemical reactions, particulate matter dynamics, acid deposition processes, and visibility considerations.

The CMAQ model demonstrates remarkable versatility by utilizing a generalized coordinate system, allowing for multi-scale and multi-layer grid simulations in spatial dimensions. To streamline the integration and data interchange between different modules of the model, the CMAQ model employs IO/API interfaces, thereby standardizing model input and output data.

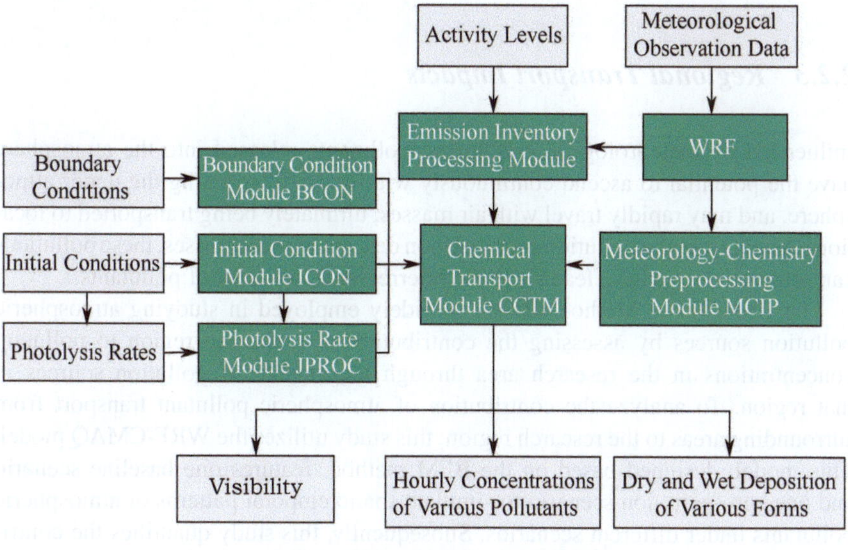

Fig. 2.6 The framework of the WRF-CMAQ air quality simulation system

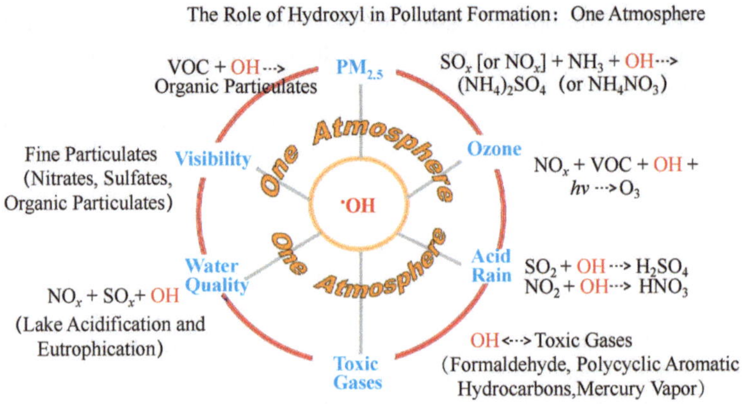

Fig. 2.7 The design concept of "One Atmosphere"

The CMAQ model is distinguished by its adaptable modular structure, affording users the flexibility to combine, modify, and replace modules as required. It offers a range of parameterization schemes for significant atmospheric physical and chemical processes, empowering users to choose appropriate chemical models aligned with their research objectives. For instance, in CMAQ v5.0.2, users can select from options such as cb05cl, cb05cl_ae5, saprc99, and saprc99_ae5 for the chemical module, as well as aero4, aero5, and aero6 for the aerosol module. This design grants users the latitude to modify the mechanisms or algorithms governing atmospheric physical and chemical processes or introduce new process modules, thereby establishing a robust platform for investigating atmospheric chemical reaction mechanisms.

2.2.3 Regional Transport Impacts

Influenced by meteorological conditions, pollutants released into the atmosphere have the potential to ascend continuously with updrafts, reaching the upper atmosphere, and may rapidly travel with air masses, ultimately being transported to locations distant from the pollution source. Upon descent with air masses, these pollutants can settle on the surface, leading to the interregional transport of pollutants.

The Brute Force Method (BFM) is widely employed in studying atmospheric pollution sources by assessing the contribution of a specific region to pollutant concentrations in the research area through the closure of pollution sources in that region. To analyze the contribution of atmospheric pollutant transport from surrounding areas to the research region, this study utilizes the WRF-CMAQ model. This model, designed based on the BFM method, features one baseline scenario and one zero-emission scenario to simulate spatiotemporal patterns of atmospheric pollutants under different scenarios. Subsequently, this study quantifies the contribution of regional transport to pollutant concentrations by evaluating the difference between sensitivity experiments with emission closure (or perturbation) and the baseline experiment. This approach allows for a quantitative assessment of the impact of major atmospheric pollutant transport from surrounding areas on the concentrations in the research region. To expound further, the baseline scenario incorporates actual emissions from the research area in the base or target year. In the zero-emission scenario, all emissions within the research area are set to zero, while maintaining other settings identical to the baseline scenario. Simulation results from the zero-emission scenario effectively illustrate the influence of pollutant transport from surrounding areas to the research region.

2.2.4 Atmospheric Environmental Capacity

Atmospheric environmental capacity denotes the utmost permissible emissions of atmospheric pollutants within a given region, aligned with specific environmental objectives, such as air quality compliance or critical loads for acid deposition. It serves as a pivotal foundation for regulating the overall emissions of air pollutants and overseeing air quality. This concept represents the maximum threshold capacity of a region's atmospheric environment system concerning the developmental scale of the region and various activity elements. In contrast to atmospheric environmental carrying capacity, atmospheric environmental capacity assumes a more direct and specific role in guiding the control of total atmospheric pollutant emissions and the improvement of ambient air quality.

In China, the comprehensive strategy for controlling total atmospheric pollutant emissions can be encapsulated in several key approaches:

1. Total Ideal Atmospheric Environmental Capacity Control: The intricate self-purification processes within the atmosphere are characterized by complex physical and chemical interactions. Some facets of atmospheric self-purification are currently at the forefront of scientific inquiry, introducing significant uncertainties in delineating the ideal atmospheric environmental capacity.
2. Total Target Atmospheric Environmental Capacity Control: In regional assessments, a widely employed strategy involves using ambient air quality goals or corresponding standards as the foundation for determining atmospheric environmental capacity. This approach entails restricting total pollutant emissions for a given region to the maximum allowable load, ensuring compliance with established ambient air quality standards.
3. Total Directive Atmospheric Environmental Capacity Control: This approach involves the issuance of primary pollutant emission control targets by national and local governments within a specified timeframe, guided by established principles. Referred to as total directive atmospheric environmental capacity control, this method entails the determination of reasonable allocation rates for various sub-regions within the total control target range. Typically, the allocation rates are determined through comparative analysis and proportional allocation methods,

considering the proportional relationships of representative indicators such as regional social, economic, resource, and area factors. The advantages of total directive atmospheric environmental capacity control lie in its simplicity, practicality, and rapid effectiveness. However, its limitation stems from the absence of a direct correlation between controlled total pollutant emissions and ambient air quality, posing challenges in understanding how to mitigate pollution sources and determining the necessary reductions to achieve environmental air quality standards.

In adherence to the criteria delineating conformity with ambient air quality standards, the regulation of atmospheric environmental capacity manifests through diverse modalities. These encompass the realization of compliance predicated on city-wide average values, the fulfillment of regulatory requirements at all designated air quality monitoring points, adherence to prescribed annual average concentration standards, and the satisfaction of stipulated daily average concentration standards.

2.2.4.1 Traditional Estimation Methods

Conventional techniques for estimating atmospheric environmental capacity encompass the A-P value method and the linear programming method.

The A-P value method stands out as the most straightforward approach to atmospheric environmental capacity estimation. It circumvents the necessity for detailed knowledge pertaining to the configuration, emissions, and emission methods of identified pollution sources. Leveraging geographical area and natural conditions, this method offers a preliminary estimation of the atmospheric environmental capacity within the study area. While it proves to be simple and practical, providing reference value for pertinent policy formulation and regional total control indicators, the A-P value method grapples with the challenge of adequately incorporating the influence of external pollutants and local dust sources on the environment. Typically treating external pollutants and local dust sources as integral to the environmental background, it falls short in estimating the maximum permissible emissions from local dust sources. Furthermore, the A-P value method imposes emission reduction without considering the distinct contributions of various pollution sources to ambient air quality. This results in a scenario where emissions reduction from certain sources, though having relatively minor impacts on ambient air quality, is disproportionately substantial due to high emission levels. Conversely, pollution sources with more substantial impacts on ambient air quality despite not having particularly high emissions, often characterized by lower vertical emission heights, tend to undergo relatively modest reductions under the A-P value method.

The linear programming method offers a more intricate portrayal of the "emission source-receptor" dynamic, allowing for simultaneous optimization of atmospheric environmental capacity allocation at a regional scale. Nonetheless, this method grapples with limitations stemming from linear response relationships, rendering it inadequate for representing the intricate non-linear processes inherent in atmospheric compound pollution.

Conventional techniques for estimating atmospheric environmental capacity are beset by several noteworthy challenges:

1. Estimation methods for the atmospheric environmental capacity of gaseous pollutants exhibit individual strengths, yet the theoretically calculated permissible emissions for each pollution source raise concerns regarding economic and technical feasibility.
2. Traditional approaches to total particulate matter control are primarily directed at organized emission sources such as smoke and industrial dust, lacking a direct correlation with ambient air quality. Consequently, estimating the atmospheric environmental capacity for particulate matter becomes impractical. Given that particulate matter in ambient air originates from diverse emission source categories, the contribution of each category to particulate matter in the ambient air consumes a portion of the atmospheric environmental capacity. Aligning with ambient air quality objectives, the implementation of total target atmospheric environmental capacity control for particulate matter necessitates a nuanced consideration of the contribution and allocation rate of each emission source category to particulate matter concentration in the ambient air. This involves determining target allowable emission levels and contributions for each category through specific technical methodologies and means. Only through such an approach can total particulate matter control be intricately linked to ambient air quality, thereby addressing the critical question of whether the total particulate matter control plan can meet ambient air quality standards. Furthermore, when executing total particulate matter control, considerations of economic and technical feasibility should encompass the reducible space for each emission source category. While transitioning to clean energy is an effective strategy for mitigating air pollution, considering China's national conditions, the predominant reliance on coal-based energy in most cities in China is unlikely to undergo fundamental changes in the foreseeable future. Among all particulate matter emission source categories, coal smoke and dust are intricately linked to social and economic development. Typically characterized by a relatively high control level, the reducible space is limited, preventing equal treatment with other emission source categories, i.e., coal smoke and dust cannot be curtailed proportionally to other emission source categories. Emphasis should be placed on maximizing reductions within feasible constraints while sustaining economic development. In contrast, open emission source categories exhibit greater reducible space and comparatively lower control costs. Prioritizing the reduction of dust emissions proves more environmentally beneficial, warranting a strategic focus on mitigating dust sources.
3. Traditional approaches to total atmospheric pollutant control focus exclusively on calculating atmospheric environmental capacity for primary pollutants,

neglecting the intricate interplay between gaseous precursors and secondary particulate matter. Nevertheless, secondary particulate matter significantly influences $PM_{2.5}$ concentrations and cannot be dismissed as negligible.

The estimation of urban atmospheric environmental capacity, encompassing spatial extent, pollution source emissions, and pollutant transport and transformation, exhibits the following principal characteristics:

1. The regional spatial scale is extensive, spanning from tens to several hundreds of kilometers, featuring intricate underlying surfaces and meteorological conditions.
2. Emission sources are varied, unevenly distributed, and numerous in nature.
3. Atmospheric pollution manifests regional and compound characteristics, with particular emphasis on the significance of the secondary pollution process that cannot be overlooked.

In light of the aforementioned characteristics, the application of traditional methods for estimating atmospheric environmental capacity, including the A-P value method and linear programming method, introduces notable uncertainty.

2.2.4.2 WRF-CMAQ Method

To address the various limitations inherent in traditional methods of atmospheric environmental capacity estimation, this study adopts an innovative approach. Utilizing the WRF-CMAQ air quality model and an urban air pollution emission inventory, an iterative algorithm is employed for atmospheric environmental capacity with the explicit constraint of attaining planned ambient air quality objectives. This method takes into account the impact of both meteorological conditions and anthropogenic factors. Leveraging the model's simulation capabilities for intricate physical and chemical processes in the atmosphere, the study independently simulates the maximum allowable emissions for urban SO_2, NOx, primary $PM_{2.5}$, NH_3, and VOCs. These simulations are conducted with the ultimate aim of providing a scientific foundation for the formulation of urban air pollution control strategies, aligning with the objective of achieving an annual average concentration of 35.0 $\mu g/m^3$ for urban $PM_{2.5}$. The specific methodologies employed are as follows:

1. Simulation of Annual Average $PM_{2.5}$ Concentrations under the Baseline Scenario: An air quality simulation system tailored for urban settings is established using the WRF-CMAQ model. This system is employed to simulate the average concentrations of urban $PM_{2.5}$ and key components during the months of January, April, July, and October for the baseline year.
2. Setting of $PM_{2.5}$ Compliance Limits: The annual average compliance limits are derived based on the phased $PM_{2.5}$ targets stipulated in ambient air quality planning. This determination takes into account the average proportions of urban sulfate, nitrate, ammonium, and primary $PM_{2.5}$ to total $PM_{2.5}$, as simulated in January, April, July, and October of the baseline year. The established annual

average compliance limits encompass $PM_{2.5}$, sulfate, nitrate, ammonium, and primary $PM_{2.5}$.

3. Formulation of Reduction Plans: Assuming a linear relationship between pollutant emissions and $PM_{2.5}$ concentrations within a defined emission reduction range, reduction plans are devised for SO_2, NO_x, NH_3, and primary $PM_{2.5}$. These plans are based on the ratios between the annual average concentrations of sulfate, nitrate, ammonium, and primary $PM_{2.5}$, and the compliance limits. Additionally, pollution source emission inventories and particle source apportionment results are considered.
4. Computation of a new multi-pollutant emission inventory is conducted in alignment with the formulated reduction plans. Subsequently, the air quality simulation system is utilized to simulate the annual average concentrations of urban $PM_{2.5}$ and key components under the new reduction plans. Subsequently, steps (3) and (4) are iteratively repeated until the annual average $PM_{2.5}$ concentration meets the specified standard. The outcome is the determination of the maximum allowable emissions for SO_2, NO_x, and primary $PM_{2.5}$, signifying the atmospheric environmental capacity under the constraint of the targeted $PM_{2.5}$ concentration.

Factors influencing the estimation of atmospheric environmental capacity encompass anthropogenic and natural elements. Anthropogenic factors include ambient air quality standards, pollution source emission inventories, pollution causes, and the influence of external pollution source transport. Natural factors comprise meteorological conditions, topography, and pollutant background concentrations. Consequently, in the estimation of atmospheric environmental capacity, this study is grounded on the following assumptions:

1. No Significant Changes in Pollution Source Layout: This study assumes that the layout of pollution sources within the study area remains relatively stable in the target year, with no noteworthy alterations.
2. No Significant Changes in Meteorological Conditions: Meteorological factors, including wind speed, wind direction, mixing layer height, and precipitation, may exhibit variations in different years. This study relies on meteorological conditions from the baseline year for estimation, assuming no significant changes in these conditions.
3. Proportional Reductions in Emissions for Surrounding Cities: Recognizing the non-negligible impact of regional transport on achieving ambient air quality standards, accounting for reductions in pollution from surrounding areas not only mirrors real-world scenarios but also enhances the overall credibility of emissions reduction efforts.

2.3 Measures Assessment

2.3.1 Overview of Measures Assessment

The swift progress of industrialization and transportation has hastened the urbanization process, resulting in the widespread consumption of fossil fuels like coal, oil, and natural gas. This has, in turn, given rise to atmospheric environmental pollution. Since 2001, the World Meteorological Organization (WMO) has been actively engaged in researching urban air pollution and assessing the impact of diverse emission control strategies. Existing studies predominantly delve into the effects of air pollution control policies on ambient air quality, examining both quantitative and qualitative aspects.

Qualitative analysis involves the process of acquiring necessary pollutant concentration data or other indicators characterizing ambient air quality through observations or literature research. This method includes comparing variations in pollutant concentrations or ambient air quality indicators before and after the implementation of air pollution control policies, enabling a concise and intuitive comparative analysis. However, qualitative analysis does not comprehensively account for the effects of the same air pollution control policies on multiple pollutants, the coordinated impacts of different air pollution control policies on pollutants, discrepancies caused by changes in meteorological conditions, and other factors influencing ambient air quality indicators. Consequently, outcomes from qualitative analysis often carry a higher degree of uncertainty.

Quantitative analysis involves the study of the effects of air pollution control policy implementation in a quantitative manner, utilizing statistical analysis methods based on regression models, the emission inventory method grounded in actual activity levels, and the numerical simulation method based on air quality models. In quantitative research employing statistical analysis based on regression models, researchers commonly utilize regression analysis and econometric modeling methods. Ordinary least squares (OLS) regression and difference in difference (DID) analysis are frequently employed techniques. OLS regression can analyze higher powers or interactive effects between factors but may overlook the impacts of other pollution control policies and neglect endogeneity issues. DID analysis necessitates a degree of similarity between the treatment and control groups in all aspects other than the pollution control policy under study; otherwise, biased regression results may occur. Thus, the careful selection of the control group is crucial.

In light of the continuous enrichment of foundational data and ongoing enhancements in relevant models, scholars have employed emission inventory models and air quality models to evaluate the ecological advantages resulting from air pollution control measures, culminating in the establishment of a sophisticated methodology. The methodology for evaluating the environmental benefits of air pollution control measures can be succinctly outlined as a quantitative analysis of the impact of historical changes across various dimensions on regional atmospheric emissions and their concentrations. This analysis is conducted by maintaining factors such as energy consumption, product output, and control technology at the historical period's level. When employing an air quality model to simulate the contributions of air pollution control measures to the amelioration of ambient air quality, the precision of the assessment outcomes is intricately tied to the dependability of the emission inventory.

In the current investigation, the urban-scale emission inventories for greenhouse gases and atmospheric pollutants, as delineated in Sect. 2.1, are employed to apprehend the dynamic effects of diverse atmospheric pollution control measures on emissions originating from distinct source categories.

2.3.2 Classification of Measures

To achieve the objective of "dual attainments", China has implemented a diverse array of air pollution control policies at both the national and provincial levels. These policies encompass areas such as energy development, energy efficiency, green transportation, low-carbon construction, resource recycling, carbon sequestration enhancement, market finance, regulatory development, and international cooperation. This study systematically reviews pertinent policies, emphasizing four key dimensions: energy structure, industrial structure, transportation structure, and land use structure. The goal is to elucidate the primary objectives and concentrate on pivotal areas. Fig. 2.8 provides a summary of the principal measures aimed at reducing pollutant and carbon emissions within the context of these four restructuring frameworks.

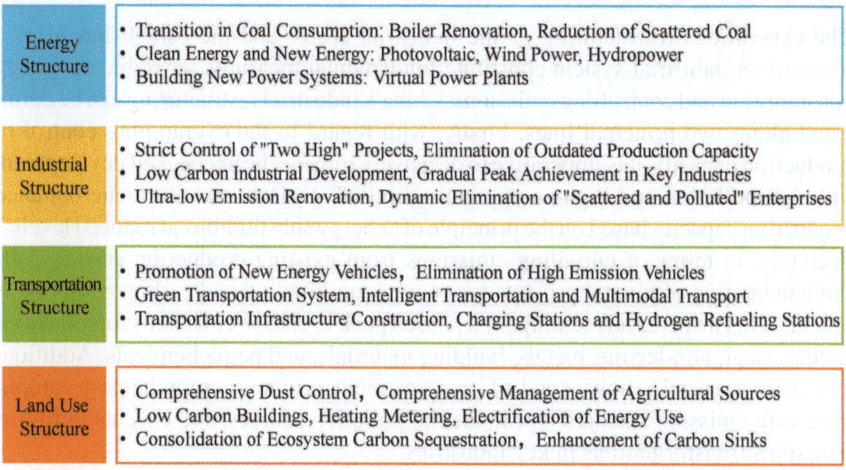

Fig. 2.8 The principal measures for reducing pollution and carbon emissions within the context of the four restructuring frameworks

2.3.2.1 Energy Structure

The restructuring of the energy sector stands as a pivotal imperative for China's energy development, crucial for ensuring energy security and attaining the urban "dual attainments" objective. The primary thrust of energy restructuring is centered on diminishing the demand for and consumption of fossil fuels, reducing the proportion of coal-fired power generation, and actively advancing new and renewable energy sources.

Throughout the 13th Five-Year Plan period, China implemented a series of green development measures to optimize and adjust the energy structure. Notable efforts were directed towards the comprehensive remediation of coal-fired boilers. Consequently, coal-fired boilers with a capacity of 35 steamed tons per hour or less were virtually eradicated in key regions, and in urban areas at the county level or above nationwide, coal-fired boilers with a capacity of 10 steamed tons per hour or less were effectively eliminated. Robust initiatives were undertaken to promote clean heating in northern regions, with central government financial support allocated to pilot cities in the northern regions for winter clean heating. This initiative covered the Beijing-Tianjin-Hebei region and surrounding areas, as well as the Fenwei Plain, leading to the replacement of bulk coal for approximately 25 million households. The implementation of these measures has successfully reduced the reliance on fossil fuels, particularly coal, and has fostered the development of natural gas, non-fossil energy sources, new energy, and renewable energy, yielding commendable outcomes.

2.3.2.2 Industrial Structure

The expeditious restructuring of the industrial sector and the establishment of a low-carbon industrial system constitute fundamental measures for reducing energy consumption and controlling emissions. China's industrial restructuring can be delineated along two principal lines. Firstly, with regard to the overarching control of production capacity, it is imperative to decisively impede the unchecked development of "high-pollution and high-energy consumption" projects and eliminate outdated production capacity based on the principle of "energy substitutions at reduced levels". Secondly, in terms of curtailing emissions from existing production capacity, the implementation of pivotal projects for energy conservation and carbon reduction is crucial. This involves advocating for a phased peak in carbon emissions for industries such as steel, non-ferrous metals, building materials, and petrochemicals. Additionally, there is a need to bolster end-of-pipe treatment, promote comprehensive compliance with emission standards for industrial pollution sources, and execute ultra-low emissions transformations in key industries.

Over the 13th Five-Year Plan period, China achieved notable success in industrial restructuring. In the realm of capacity control, proactive measures were taken to address overcapacity in industries like steel, coal, coal-fired power, and cement, with a specific focus on achieving the dynamic clearance of "scattered, non-compliant, and polluting" sources in key areas. In terms of end-of-pipe treatment, the accumulated installed capacity of ultra-low emission coal-fired power units nationwide reached 950 million kilowatts, the capacity of ultra-low emissions transformation in the steel industry reached 620 million tons, and more than 70,000 industrial furnaces and kilns underwent treatment.

2.3.2.3 Transportation Structure

The restructuring of the transportation sector in China can be categorized into two primary dimensions. The first dimension involves the low-carbon transformation of transportation vehicles, encompassing the elevation of emission standards, the gradual phasing out of outdated vehicles and machinery, the promotion of new energy sources, and the development of supporting infrastructure such as charging stations and hydrogen refueling stations. The second dimension focuses on the green and efficient integration of transportation methods, emphasizing the development of smart transportation and multimodal integration to augment the proportion of environmentally friendly urban travel.

Throughout the 13th Five-Year Plan period, China undertook concerted pollution control initiatives related to "oil, roads, and vehicles", sustaining efforts to promote the restructuring of transportation. Regarding the low-carbon transformation of transport vehicles, more than 14 million outdated motor vehicles were phased out since the end of 2015, and the percentage of new energy buses increased from 20% to over 60%. As of July 1, 2020, the nationwide implementation of China VI emission standards for light-duty vehicles was initiated, accompanied by the comprehensive supply of China VI standard automotive gasoline and diesel fuels. This achievement marked the convergence of automotive diesel, regular diesel, and specific ship fuels. In the realm of transportation restructuring, a total of 92 dedicated railway lines were completed or opened nationwide in 2020, contributing to a more than 20% increase in the national railway freight volume compared to 2017.

2.3.2.4 Land Use Structure

The restructuring of land use in China encompasses three principal dimensions. The first dimension prioritizes pollution control, particularly focusing on the comprehensive management of dust and agricultural sources. This entails emission reductions in particulate matter and ammonia, both of which exhibit poor synergy. The second dimension centers on optimizing construction practices by promoting low-carbon buildings and electrifying building energy use. The third dimension involves carbon sink management, aiming to enhance the carbon sequestration capacity of ecosystems through initiatives such as the Grain for Green program and afforestation.

Throughout the 13th Five-Year Plan period, China successfully reduced pollutant emissions via land use restructuring. In terms of dust source control, continuous efforts were directed towards achieving "six 100% compliance" at urban construction sites, with over 230,000 construction sites nationwide participating in dust control measures. In key regions, the mechanical sweeping rate for urban roads surpassed 90%. Regarding agricultural source management, there was a gradual increase in the comprehensive utilization of straw, and stricter supervision was enforced to curb open-field straw burning. In 2020, the number of straw-burning sites nationwide decreased by 30.5% compared to 2017. Furthermore, the proportion of newly constructed green buildings in urban areas reached 77% in 2020.

2.3.3 Evaluation Methods

In recent years, cities across China have undertaken various initiatives to enhance ambient air quality and reduce carbon emissions, with a focus on restructuring, end-of-pipe treatments, and urban management.

This study has systematically collected and organized relevant policy documents related to the "dual attainments" in China, including the "Ten Measures for Air", the "Three-Year Action Plan", and policies addressing clean heating transformation in northern regions and ultra-low emissions transformation in the steel industry. Through a detailed organization of policy content, implementation specifics, and target responsibility documents for specific regions, this study has extracted measure details and quantitative parameters from four main aspects: energy restructuring (such as power restructuring and the transition to cleaner civil fuels), industrial restructuring (including the elimination of outdated production capacity, rectification of "scattered, non-compliant, and polluting" enterprises, transformation of key industries toward ultra-low emissions, and upgrading of end-of-pipe industrial treatment), transportation restructuring (covering the shift from roads to railways, transportation cleanliness, upgrading vehicle emission standards, and phasing out old vehicles), and land use restructuring (comprehensive treatment of agricultural and dust sources). These initiatives can be summarized as nine key actions: power restructuring and deep governance, clean fuel substitution, comprehensive remediation of boilers, dissolution of excess production capacity, industrial upgrading and renovation, management of volatile organic compounds, prevention and control of mobile source pollution, comprehensive treatment of dust sources, and management of agricultural non-point sources.

Utilizing the city's emission inventory from the baseline year and considering atmospheric environmental capacity, socioeconomic development, and policy planning status, a thorough analysis is conducted for each specific measure. This includes quantifying their effects on energy consumption data, energy efficiency, industrial product output, technology distribution, and the extent of end-of-pipe treatment. Additionally, the energy consumption and pollutant emissions under different pollution control measures and combinations of measures are quantified through the integration of multi-source data and coupling of multi-level models. The detailed technical roadmap is depicted in Fig. 2.9.

The methodology employed to quantify reductions in emissions and contributions resulting from diverse pollution control measures is delineated as follows:

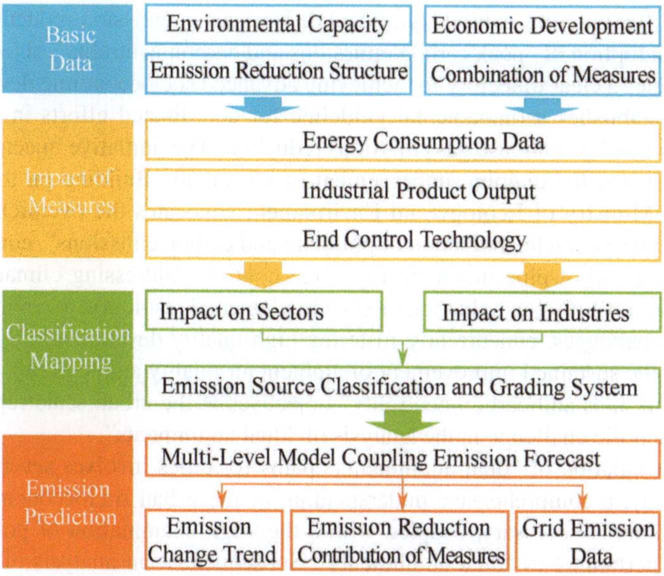

Fig. 2.9 The technical roadmap for quantifying the effectiveness of pollution control measures

1. Assuming non-implementation of the pollution control measure by the target year, the energy efficiency, technology proficiency, and end-of-pipe treatment standards associated with this measure will persist at the baseline year's status.
2. Building upon this assumption, activity levels and emission factors for pertinent emission source categories in the target year are extrapolated. Consequently, the emissions of atmospheric pollutants and greenhouse gases are computed, depicting the emissions in an uncontrolled scenario for the target year, where the pollution control measure has not been enacted.
3. Through a comparative analysis between the emissions in the uncontrolled scenario for the target year and the actual emissions, the magnitude of emission reductions and the contribution of the pollution control measure can be assessed. This analytical framework forms the basis for subsequent evaluations of air quality improvement and cost considerations.

2.4 "Dual Attainments" Analysis

Overview of "Dual Attainments" AnalysisCities are the fundamental units for assessing national air quality management, constituting the primary front for combatting air pollution. With more than half of China's population residing in cities, urban centers consume over 80% of the nation's energy, resulting in more than 80% of carbon emissions and over 60% of atmospheric pollutant emissions. Due to distinct levels of economic development, diverse industrial structures, and varying emission sources, different cities exhibit significant differences in emission reduction potentials. The adoption of "peak carbon emissions and carbon neutrality" objectives not only provides a clear trajectory for achieving advanced socioeconomic development but also establishes a fundamental guideline for coordinated efforts in air pollution control and greenhouse gas emissions reduction. This initiative injects renewed momentum into the ongoing improvement of air quality. Building on this framework, the Ministry of Ecology and Environment advocates the overarching principle of "synergizing the reduction of pollutant and carbon emissions", emphasizing a shift from end-of-pipe treatment to source control. Addressing climate change through carbon emission reduction offers a fundamental resolution to environmental pollution challenges, concurrently fostering high-quality development. Effectively achieving the sustained improvement of ambient air quality and realizing the "peak carbon emissions and carbon neutrality" objectives at the urban scale represents a crucial scientific challenge in the analysis of "dual attainments".

The formulation of "dual attainments" paths for cities involves several crucial steps. Firstly, a comprehensive understanding of the urban socioeconomic development status is necessary, coupled with a thorough examination of policies and measures pertaining to "dual attainments". Subsequently, an analysis of emission trends, structures, and distribution is conducted, utilizing urban-level emission inventories for greenhouse gases and atmospheric pollution sources. This analysis aims to identify key emission reduction areas and regions within the city. Moreover, constrained by air quality compliance, the calculation of atmospheric environmental capacity is essential, alongside the prediction of future energy consumption and pollution emissions, referencing relevant coordinated emission reduction plans. Building upon this foundation, proposed paths to achieve urban air quality standards are delineated, and the peak status of carbon emissions is analyzed after verifying the feasibility of the targets. Lastly, the urban "dual attainments" paths are formulated by integrating the analysis results of air quality compliance and peak carbon emissions.

When devising "dual attainments" paths for cities, it is imperative to focus on control areas, giving particular attention to hot spots characterized by high concentrations of both atmospheric pollutants and greenhouse gas emissions. Additionally, regions in close proximity to high-emission sources and with dense populations should be considered as priority areas for coordinated control. Regarding the selection of controlling objects, each city should prioritize industries with high energy consumption and emissions, aligning with their unique industrial structures.

The assessment of "dual attainments" paths for cities encompasses various dimensions, with a central focus on the prediction of the reduction in atmospheric pollutants and CO_2 emissions, along with air quality compliance. Starting from the emission inventory in the baseline year, the methodology for setting emission reduction scenarios, predicting future emissions, and evaluating prospective air quality improvements, in conjunction with air quality models, has matured significantly, primarily applied at the national and provincial levels in the past. This study extends the application of such methodology to the urban scale.

2.4.1 Analysis Methods for "Dual Attainments"

The urban "dual attainments" coordinated analysis framework is elucidated in Fig. 2.10. This comprehensive framework integrates emission inventories, numerical simulations, and measure assessments. It facilitates the calculation of energy consumption and emissions under policy and technological interventions, enabling an analysis of urban air quality and carbon emission changes in the target year. The framework comprises four main components:

1. Iterative calculation of the maximum allowable emissions for each pollutant is conducted based on urban inventories and model simulations, with a primary focus on achieving urban air quality standards.
2. Utilizing the emission source analysis results, urban emission reduction potential, and the formulation of emission reduction plans aligned with city planning, future energy and technology scenarios, along with their corresponding emissions, are derived.
3. After confirming the feasibility of the targets, the paths to achieve urban air quality standards are determined. Simultaneously, through the analysis of carbon emissions in this scenario, the situation of urban peak carbon emissions under the guidance of air quality goals can be obtained.
4. By comparing the baseline scenario with the policy scenario, the effects of different energy policies, industrial structures, and technological applications on greenhouse gas and atmospheric pollutant emissions are analyzed.

The subsequent sections delineate the methodologies within the "dual attainments" framework for analyzing the paths toward urban air quality compliance and peak carbon emissions.

2.4.2 Analysis of the Paths to Urban Air Quality Compliance

The technical methodology employed for analyzing the paths toward urban air quality compliance primarily focuses on comprehending the causes of atmospheric pollution and establishing the correlation between pollutant emissions and ambient air quality. Initially, by collecting and scrutinizing urban air quality data and pollution sources, a thorough analysis of pollution issues and their origins is conducted. This analysis, combined with projections of socioeconomic and energy development, as well as the air quality enhancement requirements for compliance, aids in determining the timeline for achieving urban air quality compliance and setting interim air quality objectives, thereby delineating the compliance pressures. Subsequently, crucial issues pertaining to meeting urban air quality standards are pinpointed, including understanding regional transport patterns, identifying critical pollution contributors, recognizing key industry enterprises, and assessing spatial sensitivity. Additionally, a practical pollution control plan is devised, taking into account factors such as industrial

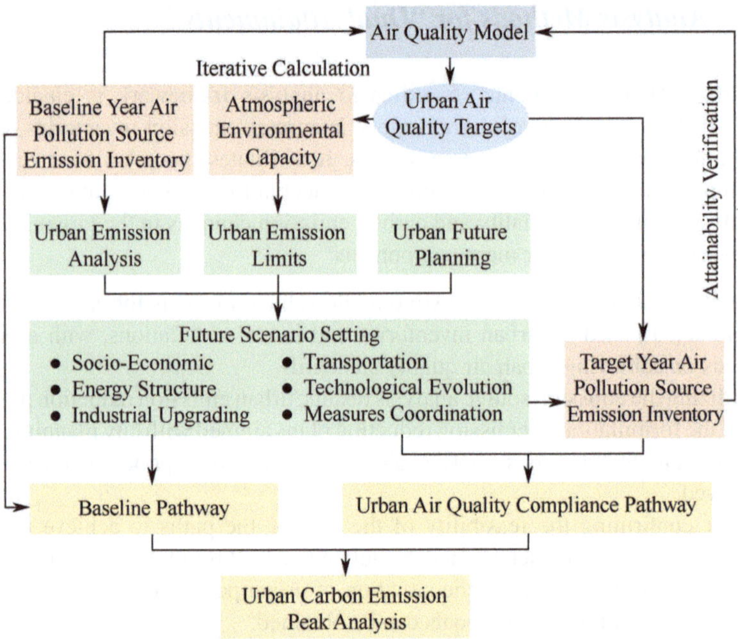

Fig. 2.10 The urban "Dual Attainments" coordinated analysis framework

structure, energy composition, transportation layout, and land use configuration. The potential for reducing pollutant emissions through each measure is evaluated. Finally, air quality models are employed to simulate the anticipated effects on air quality resulting from the implementation of this pollution control plan. If the urban air quality objectives are attainable, the pollution control plan becomes the optimal strategy for improving urban air quality. Conversely, if the goals are not met, the measures and scenarios are adjusted iteratively to eventually arrive at the optimal compliance scenario. The technical roadmap for urban air quality compliance is illustrated in Fig. 2.11.

In overview, the examination of pathways toward urban air quality compliance encompasses seven key facets. These include the analysis of major atmospheric pollution issues and their causes, the identification of key issues in air quality compliance, the research on air quality goals and compliance deadlines, the calculation of atmospheric environmental capacity, the analysis of emission reduction potential, the formulation of air pollution prevention and control plans, and the analysis of the reachability of planning objectives.

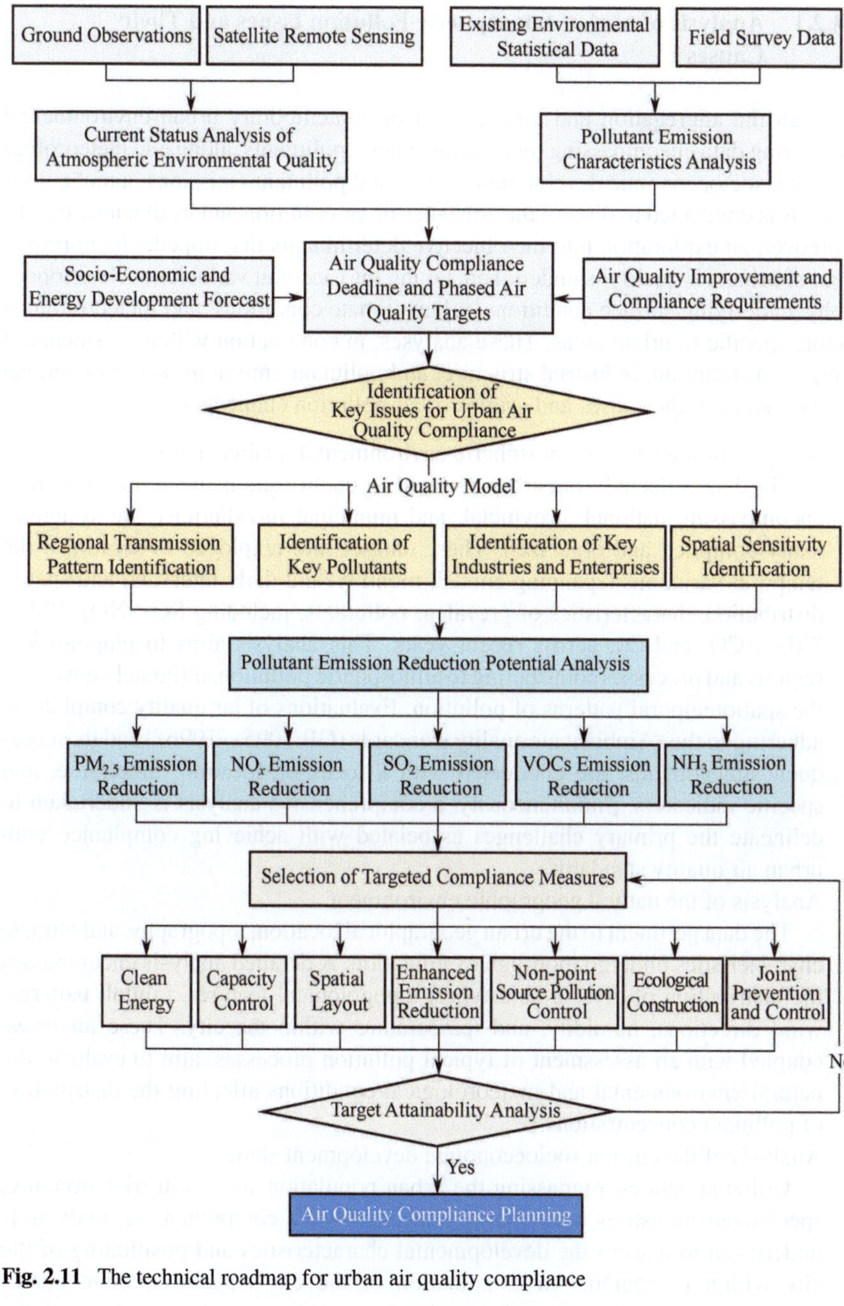

Fig. 2.11 The technical roadmap for urban air quality compliance

2.4.2.1 Analysis of Major Atmospheric Pollution Issues and Their Causes

Through the aggregation and arrangement of contemporary urban environmental monitoring data encompassing major atmospheric pollutants, alongside meteorological data, socioeconomic development status, and pollutant emissions, a meticulous analysis is conducted to discern the attributes of air pollution and its dynamic trends. Moreover, an exploration into the objective determinants that impede the improvement of urban air quality is undertaken, taking into account variables such as topography, underlying surface conditions, urban climate conditions, and meteorological factors specific to urban areas. These analyses, in conjunction with assessments of energy consumption, industrial structure, and pollutant emissions, serve to unravel the intricacies of the causes and origins of air pollution challenges.

1. Analysis of the current atmospheric environmental quality status

   The data collected from urban ambient air quality automatic monitoring sites, encompassing national, provincial, and municipal jurisdictions, are systematically compiled and organized. These datasets are employed to scrutinize the temporal variations (spanning annual, monthly, and daily intervals) and spatial distribution characteristics of prevalent pollutants, including SO_2, NO_2, PM_{10}, $PM_{2.5}$, CO, and O_3, across recent years. This analysis aims to pinpoint key regions and processes contributing to atmospheric pollution, ultimately unveiling the spatiotemporal patterns of pollution. Evaluations of air quality compliance, adhering to the "Ambient air quality standards (GB 3095—1996)" and its subsequent amendments, are conducted with a focus on location, timeframe, and specific indicators. Simultaneously, a comprehensive analysis is undertaken to delineate the primary challenges associated with achieving compliance with urban air quality standards.
2. Analysis of the natural geographic environment

   The data pertinent to the urban geographical location, topography, and climate characteristics undergo thorough examination. A detailed analysis encompasses the distribution of diverse land types, topographical features, rainfall patterns, wind directions, humidity, and temperature within the city. These analyses, coupled with an assessment of typical pollution processes, aim to evaluate the natural environmental and meteorological conditions affecting the distribution of pollutant concentrations.
3. Analysis of the current socioeconomic development status

   Utilizing data encompassing the urban population size, industrial structure, specialized industries, and energy consumption, a comprehensive analysis is undertaken to discern the developmental characteristics and positioning of the city within its region. An examination of the city's socioeconomic conditions in different districts and counties is carried out, and an analysis of the economic driving factors influencing the urban air pollution status and developmental trends is conducted. This analysis is approached from the perspectives of economic development, energy consumption, transportation development, and urbanization.

4. Analysis of pollutant emission characteristics

   Based on the established emission inventory of urban air pollution sources, an analysis is undertaken to ascertain the contribution rates of various emission sources, including urban specialized industries, motor vehicles, dust sources, and living non-point sources, to the emissions of major pollutants such as $PM_{2.5}$, PM_{10}, SO_2, NO_x, and VOCs. The sources of atmospheric pollutants are also scrutinized. Moreover, different regions, industries (or emission source categories), energy utilization methods, pollution discharge facilities, and regulatory levels are taken into consideration to summarize the emission characteristics of major atmospheric pollutants. Building upon this foundation, key emission areas (or units), types of pollution sources, and emission stages are identified for different seasons.

2.4.2.2 Identification of Key Issues in Air Quality Compliance

1. Regional transport analysis

   Investigations are conducted to assess the impact of external emission sources on urban pollutants. Specifically, within the urban context, the WRF-CMAQ model is utilized (detailed methods outlined in Sect. 2.2.3) to analyze the characteristics of regional atmospheric pollutant transport flow fields. Moreover, the model is employed to quantitatively simulate the contributions of both local and external emission sources to the annual average concentrations of PM_{10} and $PM_{2.5}$, severe pollution periods, and excess O_3 hours across the entire city and various air quality monitoring locations. Subsequently, the causes of severe pollution episodes within urban areas and the locations of external emission sources are subjected to further analysis.

2. Sensitivity analysis

   Utilizing modeling tools, the city is subdivided into multiple spatial units or equidistant grids. The influence of pollutant emissions on urban air quality is individually simulated for each spatial unit, employing the backward trajectory model. The environmental sensitivity of different spatial units is assessed. Taking into account meteorological flow fields, underlying surfaces, and topography, a comprehensive analysis is then conducted on the distinct characteristics of atmospheric environmental sensitivity in various spatial units. This analysis serves as a foundation for optimizing industrial layout and developing spatially differentiated control strategies.

2.4.3 Research on Air Quality Goals and Compliance Deadlines

1. Economic energy consumption forecast

Information related to urban planning and layout, economic growth (regional GDP and per capita income), energy consumption, and energy structure (projected through industrial development planning) is assessed and gathered. Additional data on population growth, changes in industrial structure, motor vehicle ownership and usage, and the expansion (or contraction) of the electricity sector and industry are also collected. These datasets are employed to forecast future socioeconomic development parameters such as GDP, energy consumption, coal consumption, electricity-related coal consumption, motor vehicle ownership, and major industrial product outputs. This forms the basis for predicting atmospheric pollutant emissions.

2. Analysis of compliance gap

   Predictions of pollutant emissions for the next 5, 10, and 15 years are made based on current policies and measures, considering the potential addition of new control measures. Subsequently, a fuzzy response relationship is established between atmospheric pollutant emissions (reductions) and ambient air quality indicators (such as $PM_{2.5}$). This is accomplished using emission inventories and pollutant concentrations, representing the impact factors of atmospheric pollutants on ambient air quality. Building upon this, a baseline scenario is constructed, encompassing pollutant emissions, concentrations, and air quality conditions. An analysis is then conducted to identify disparities between atmospheric pollutant concentrations in the baseline scenario and the air quality targets at various stages.

3. Determination of air quality targets

   The strategy to achieve urban air quality standards is formulated by considering the existing urban pollution scenario, incorporating information on major pollution sources and industries. Additionally, future urban development plans, including industrial, energy, transportation, and urban-rural planning, are taken into account. The target year for achieving ambient air quality standards is established in alignment with the urban air quality compliance strategy and the stipulated regulations of national and local authorities, as per the specific requirements of local governance.

4. Calculation of atmospheric environmental capacity

   Subject to the requirement of achieving planned ambient air quality targets and with a primary focus on safeguarding human health, the WRF-CMAQ model (comprehensive details provided in Sect. 2.2.4) is employed. Utilizing factors such as pollutant emission performance, the environmental contributions of diverse pollution sources, available space for reduction (emission reduction potential), and considering the influence of meteorological conditions and anthropogenic elements, this model conducts simulations of the intricate physical and chemical processes occurring in the atmosphere. Based on this framework, the maximum permissible total emissions of SO_2, NO_x, VOCs, $PM_{2.5}$, and PM_{10} are established within the confines of phased ambient air quality targets.

5. Analysis of emission reduction potential

   Aligned with national, provincial, and municipal developmental blueprints, as well as pertinent industrial strategies, the prognostication of major atmospheric pollutant emissions—comprising SO_2, NO_x, VOCs, $PM_{2.5}$, and PM_{10}—across the entire societal spectrum and prominent industries is conducted. This forecast meticulously integrates considerations for the encouragement and adoption of clean and renewable energy sources, incorporating regulatory mandates, policies, and standards governing both new and existing pollution sources. Simultaneously, an evaluation is made regarding the impact of emergent pollutant emissions on reduction initiatives, coupled with an assessment of the reduction potential inherent in energy structure, industrial configuration, transportation arrangements, and land use organization. Additionally, a detailed appraisal of emission reduction possibilities within key industries is undertaken, encompassing source control, process-based emission abatement, and end-of-pipe treatment methodologies.

6. Formulation of air pollution prevention and control plans

   Harmonizing with the overarching vision of promoting environmentally sustainable development, with a specific emphasis on environmental preservation, and in tandem with the diverse air pollution prevention and control regulations and initiatives disseminated and enacted at the national, provincial, and municipal echelons, the strategic blueprint for achieving urban air quality standards is devised. This planning framework strategically positions air pollution prevention and control as a pivotal catalyst for economic transformation and advancement. The planning scheme is systematically structured across three tiers: source control, end-of-pipe treatment, and ancillary support systems.

   (a) Source control primarily encompasses elements such as the primary trajectory and imperative parameters of urban socioeconomic development, the apt model for industrial growth, the judicious arrangement of functional zones, and the strategic guideline for optimizing energy consumption.
   (b) In the domain of end-of-pipe treatment, tailored to the production characteristics of distinct industries and the emission profiles of varied source categories, comprehensive pollution prevention and control plans and regulatory indictors that are scientifically robust and operationally robust are formulated. Considerations span technical and economic viability, safety protocols, and operational feasibility.
   (c) The support system aspect entails the establishment and execution of comprehensive support mechanisms across multiple dimensions, encompassing organizational structures, institutions, financial frameworks, technological provisions, policies, and public engagement. These systems are designed to ensure the effective realization of key construction initiatives and projects, bolstering capabilities for early warning and emergency responses to severe pollution events. This comprehensive approach aims at facilitating both short-term air quality regulation and long-term enhancement.

7. Analysis of the reachability of planning objectives

   Leveraging essential project objectives and emission reduction strategies for major atmospheric pollutants in the target year, and integrating high spatiotemporal resolution emission inventories for both the baseline year and the target year for air quality compliance, the WRF-CMAQ model with a three-layer nested technology is employed. This approach facilitates a quantitative simulation and analysis of the atmospheric environmental improvements resulting from the execution of air quality compliance measures in both the baseline and target years. Through this analysis, the attainability of the air quality goals outlined in the plan is quantitatively evaluated.

2.4.4 Analysis of the Paths to Urban Peak Carbon Emissions

The basis for projecting urban energy consumption and analyzing paths to peak carbon emissions is derived from the urban air quality compliance paths outlined in Sect. 2.4.3. Data sources for this analysis include statistical yearbooks, environmental statistics, and field research data. To scientifically evaluate the trends in mid-to long-term urban energy consumption and carbon emissions, it is crucial to employ suitable models. Commonly utilized models include the Greenhouse Gas-Air Pollution Interactions and Synergies model, the Market Allocation Model, the Long Range Energy Alternatives Planning (LEAP) model, the energy balance model, the system dynamics model, as well as neural network algorithms or optimization algorithms applied in certain studies.

Among the mentioned models and algorithms, the LEAP model stands out as a bottom-up energy model that initiates from the technical level, providing a depiction of energy consumption and carbon emissions within the study area. The LEAP model boasts a flexible data structure, encompassing various technologies and end-use details, thereby offering extensive options for parameterization. Widely employed for mid- to long-term energy supply and demand, as well as carbon emission forecasts in urban areas, the LEAP model is chosen in this study, taking into account factors such as data availability and model performance, to analyze the paths to urban peak carbon emissions.

2.4.4.1 LEAP Model

The LEAP model, developed by the Stockholm Environment Institute, conceptualizes the regional economy and social activities into sectors like transportation, construction, and industry, where the activities within each sector are simplified to involve the consumption and production of energy and materials. The computational steps of the LEAP model encompass several key phases. Initially, it necessitates the aggregation of energy requirements across various sectors, scrutinizing whether local resource production can fulfill their needs, and calculating resultant resource imports and exports to establish equilibrium in resource and energy demand, production, and import/export. Subsequently, relying on historical data from diverse economic sectors and scenario configurations, the model is pivotal in forecasting future energy demand and subsequently determining the energy structure and greenhouse gas emissions across different scenarios. Due to its remarkable modeling flexibility and robust scenario analysis capabilities, the LEAP model finds widespread application in research domains such as CO_2 emission reduction assessments and low-carbon development trajectories.

Within the LEAP model, energy consumption is computed by multiplying activity levels with energy intensity. The activity level, in this context, can be represented by physical indicators (e.g., production output or building area) or economic indicators (e.g., industry added value). The carbon emissions across various sectors are determined by their energy consumption and the corresponding CO_2 emission factors. It is crucial to note that when discussing urban peak carbon emissions, efforts to encourage cities to conserve energy and reduce indirect carbon emissions should include the consideration of indirect carbon emissions generated by externally purchased electricity, in addition to direct urban carbon emissions. The calculation of these emissions involves multiplying the purchased electricity volume by regional emission factors.

This study, shaped by the LEAP model's structure and data availability, introduces an analysis method for urban peak carbon emissions. The urban LEAP model developed in this study primarily consists of three modules: key assumptions, demands, and transformation (refer to Fig. 2.12).

1. Key assumptions

   The foundational assumptions within the urban LEAP model encompass key parameters such as per capita GDP, industrial structure, industrial added value by sectors, permanent population, urbanization rate, and per capita living area. Data for the baseline year are derived from the China City Statistical Yearbook, and future projections align with forecasts outlined in air quality planning paths.
2. Demands

   Demands within the urban LEAP model correspond to end-use energy consumption outlined in the energy balance sheet. The study places particular emphasis on the end-use energy demands across industrial, civil, public building, transportation, and agricultural sectors.

   (a) Industry

   Given the diversity of industrial sectors, this study undertakes the identification of the top ten energy-intensive industrial sectors. The selection is based on sub-sector energy consumption data sourced from the China City Statistical Yearbook. Other industrial sectors are simplified and categorized into "other industries". For specific industries like cement and steel, for which production output data are accessible, energy consumption is determined by the formula: energy consumption = industrial product output × energy consumption per unit of product output.

   (b) In cases where industries exhibit complex product categories or lack production information, calculations rely on energy consumption per unit of added value. Specifically, energy consumption is computed using the formula:

energy consumption = industrial added value × energy consumption per unit of added value.

(c) Civil sector

The civil sector's energy consumption analysis encompasses both heating and non-heating purposes. In northern regions with winter heating needs, heating energy usage is further divided into decentralized and central heating consumption. Estimation involves parameters such as the central heating rate, housing area, and unit heating energy consumption. Central heating utilizes heat as the terminal energy demand, with the heat production process calculated within the processing and conversion module. The type of terminal energy demand for decentralized heating primarily includes coal, natural gas, and electricity. Cities in the regions characterized by cold winters and hot summers do not have widespread central heating, so only a limited amount of decentralized heating is considered. Non-heating energy consumption, excluding heating, comprises various electrical appliances and is estimated based on factors like the number of appliances per hundred households, permanent population per household, and annual average energy consumption of electrical appliances. Data on annual average energy consumption of electrical appliances and heating energy consumption intensity are derived from the "Annual Report on China Building Energy Efficiency".

(d) Public building

Public buildings, encompassing office buildings, shopping malls, hospitals, schools, hotels, and others, primarily rely on electricity as their main energy source. Estimating the areas of various public buildings involves using yearbook data like student enrollment and bed capacity, in conjunction with parameters such as per capita area corresponding to different standards. Following this, energy consumption in public buildings is determined based on energy consumption per unit of area.

(e) Transportation

For calculating transportation energy consumption, the calculations rely on future motor vehicle ownership and anticipated changes in the proportion of motor vehicle types, emission standards, and fuel types. The Stock Turnover method from the LEAP model is directly applied, with the basic formula as follows:

$$\text{Fuel}_k = \sum_i \text{VP}_i \times \text{VKT}_i \times \text{FC}_{i,k} \tag{2.10}$$

where i represents the motor vehicle type, k denotes the fuel type, VP represents the motor vehicle ownership, VKT is the annual mileage traveled (in kilometers), FC denotes fuel economy, and Fuel represents fuel consumption. If the urban transportation department possesses a robust data foundation, encompassing year-by-year registration information categorized by motor vehicle types, a model can be advanced further utilizing the survival curve method to simulate the process of vehicle retirement and replacement [3]. The fuel consumption for each motor vehicle type is subsequently calculated and directly incorporated into the LEAP model in the form of energy usage.

(f) Agriculture

The computation of agricultural energy consumption relies on variations in the added value of the primary sector and energy consumption per unit of added value.

3. Transformation

The transformation module primarily addresses the production processes of energy subsequent to the computation of end-use energy demand. This module encompasses two main components for each processing and transformation process (Fig. 2.12 includes electricity generation, heat production, and coking, with the possibility of adding other processes like coal washing and oil refining based on urban conditions), namely the output energy and process modules. Several parameters can be adjusted, including transformation efficiency, the adoption ratios of future technologies, and import/export targets.

2.4.4.2 The Analysis of the Paths to Peak Carbon Emissions

Before projecting carbon emissions, it is imperative to input baseline year data and calibrate the LEAP model parameters to accurately reflect the city's energy consumption and carbon emissions.

This study delineates two scenarios: the business-as-usual (BAU) scenario and the air quality plans (AQP) scenario. The future macro-socioeconomic parameters (population, GDP, etc.) for both scenarios are configured similarly to the analysis of air quality compliance paths. However, the distinction lies in the fact that the BAU scenario excludes new air quality policies, whereas the AQP scenario incorporates the parameterization of relevant policies and measures from air quality compliance planning, generating coordinated effects on carbon emission reduction. The AQP scenario serves to forecast peak carbon emissions under future air quality targets. By contrasting these two scenarios, it becomes possible to further analyze the carbon emission reduction contributions of various sectors under air quality policies. The technical roadmap for analyzing the paths to urban peak carbon emissions is depicted in Fig. 2.13.

The carbon emissions calculated by the LEAP model are approached from an end-use perspective. For instance, carbon emissions arising from the utilization of electricity in the civil sector are assigned to the electricity production sector. To more intuitively depict the contributions of each sector, reallocation is conducted during the carbon emissions analysis for the case city, employing the following formula:

$$\text{Emis}_i = \sum_{j,k} \left(\text{DirectEmis}_{i,j} + \text{IndirecEmis}_{i,k}\right) \tag{2.11}$$

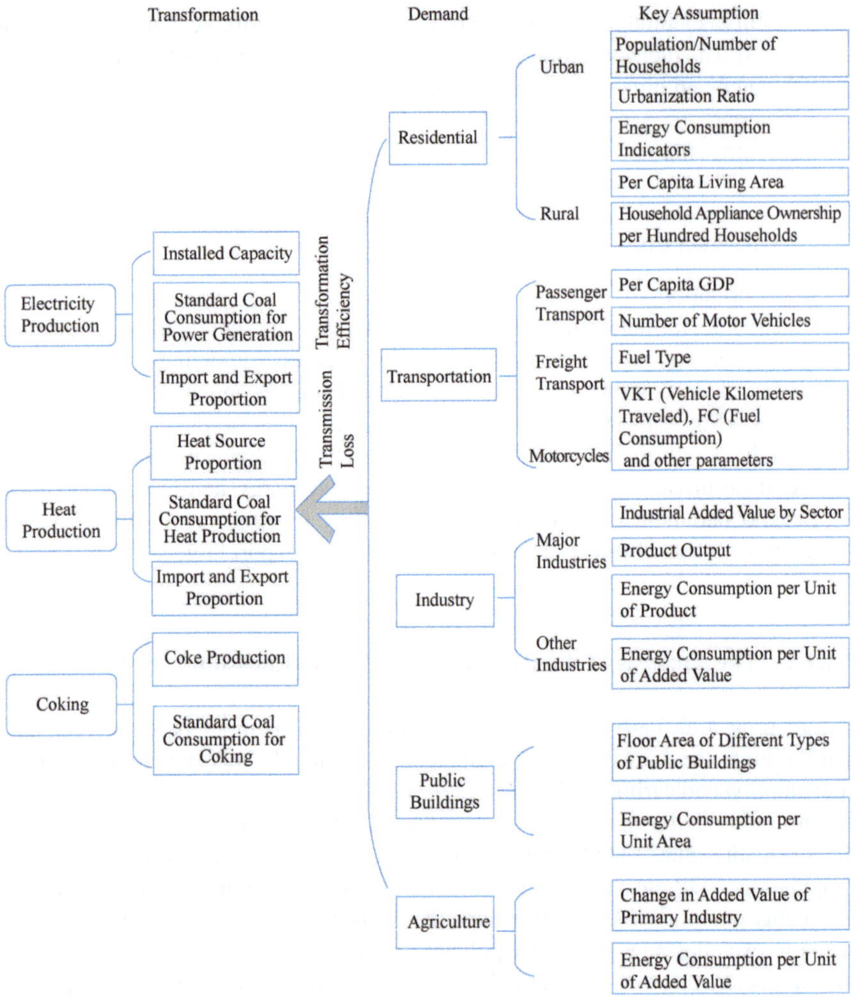

Fig. 2.12 The architecture of the urban LEAP model

$$\text{IndirecEmis}_{i,k} = \frac{\text{Demand}_{i,k}}{\text{Demand}_k} \times \text{ProEmis}_k \tag{2.12}$$

where i represents the sector, j denotes primary energy sources, such as coal and natural gas, and k represents secondary energy sources, mainly including electricity and heat. $\text{DirectEmis}_{i,k}$ and $\text{IndirecEmis}_{i,k}$ represent the direct and indirect carbon emissions from terminals, respectively. Demand refers to energy demand, and ProEmis denotes the total carbon emissions generated by the processing and conversion module for producing that fuel.

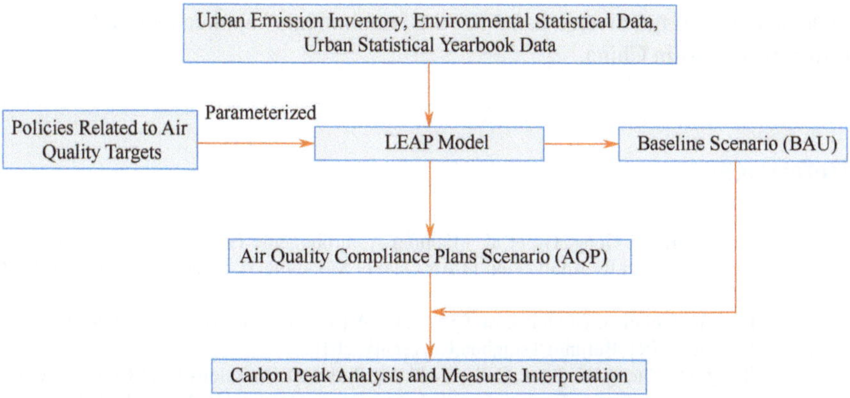

Fig. 2.13 The technical roadmap for analyzing the paths to urban peak carbon emissions

2.5 City Selection

Selected as representative case cities based on considerations of urban resource endowments, industrial structure, development positioning, regional location, and the availability of publicly quantifiable information related to socioeconomic factors, energy, low-carbon implementation plans, and air quality compliance plans, Zhengzhou City, Shijiazhuang City, and Huzhou City are examined in this study. The roles of various adjustment measures in industrial structure, energy structure, transportation structure, and land use structure are elucidated in the context of the "dual attainments" process in different cities. Additionally, common experiences are distilled, providing a robust data foundation and theoretical reference for the formulation of pertinent policies.

From a categorization standpoint, each of the three cities serves as an archetype representing a distinct city category. Zhengzhou City, positioned as a national central city, assumes a crucial role as a focal point in finance, management, culture, and transportation. Notably, it is actively establishing a contemporary and inclusive industrial system, embodying the characteristics of a comprehensive, service-oriented urban center. Shijiazhuang City, with its robust industrial foundation and a comprehensive industrial framework contributing significantly to the national GDP, exemplifies a city representative of industrial production. Huzhou City, known as the birthplace of the "Two Mountains" concept, propels its urban economic development through the cultivation of high-tech industries, rendering it a representative of cities characterized by advanced technological sectors.

Regarding geographical representation, Zhengzhou City, Shijiazhuang City, and Huzhou City stand as exemplars for the Central Plains region, the Beijing-Tianjin-Hebei region, and the Yangtze River Delta region, respectively. These cities embody the influential roles played by key city clusters in their respective regions. The exploration of the "dual attainments" trajectories in these three cities contributes to the understanding of the diverse paths to "dual attainments" applicable to cities across different regions in China.

References

1. Zheng, B., Cheng, J., Geng, G., et al. Mapping Anthropogenic Emissions in China at 1 km Spatial Resolution and Its Application in Air Quality Modeling[J]. Science Bulletin, 2021, 66, 6: 612-620.
2. He, K.B., et al. Technical Guideline on City-Level Air Pollutant Emission Inventory Development (in Chinese) [R]. Beijing: Tsinghua University, 2018.
3. Liu, J., Tong, D., Zheng, Y., et al. Carbon and Air Pollutant Emissions from China's Cement Industry 1990–2015: Trends, Evolution of Technologies, and Drivers[J]. Atmospheric Chemistry and Physics, 2021, 21: 1627-1647.
4. Liu, Z., Guan, D., Wei, W. et al. Reduced Carbon Emission Estimates from Fossil Fuel Combustion and Cement Production in China[J]. Nature, 2015, 524, 335-338.
5. Xue, W.B., Xu, Y.L., Shi, X.R. et al. Atmospheric Environment Management in China: Progress and Outlook (in Chinese) [J]. Chinese Journal of Environmental Management, 2021,13(05):52-60.

Chapter 3
Analysis of the Pathways to Achieving "Dual Attainments" in Zhengzhou City

Abstract This chapter delves into the "dual attainments" scenario within the context of comprehensive service-oriented cities, using Zhengzhou City as a case study. Through a meticulous examination encompassing the overview of Zhengzhou City, the prevailing environmental air quality conditions, the characteristics of atmospheric pollutant emissions, and the stipulated deadlines for achieving air quality standards, pathways are proposed for Zhengzhou City to attain air quality compliance based on urban emission inventories and model simulations. Simultaneously, the chapter conducts an analysis of the status of peak carbon emissions within the framework of air quality standard attainment.

According to model calculations, achieving air quality compliance necessitates significant reductions in emissions of various pollutants. Specifically, reductions of 60%, 70%, 40%, 65%, and 50% respectively for SO_2, NOx, NH_3, primary $PM_{2.5}$, and VOCs relative to their baseline levels in 2017 are deemed necessary. These reductions are projected to bring atmospheric pollutant emissions in line with the environmental capacity required for compliance. It is anticipated that by 2028, except for O_3, the concentrations of $PM_{2.5}$, PM_{10}, SO_2, CO, and NO_2 will meet the Grade II national standards for air quality. Additional coordinated measures aimed at reducing VOCs emissions are foreseen to bring O_3 concentrations into compliance by 2035. Regarding peak carbon emissions, with the target of achieving air quality compliance by 2028, carbon emissions are expected to reach their peak in 2025, after which a gradual decline is projected.

3.1 Overview of Zhengzhou City

Situated in the northern-central region of Henan Province, Zhengzhou City, the capital thereof, encompasses six districts, five cities, and one county, spanning a total area of 7446 square kilometers. As of the conclusion of 2017, the city accommodated a populace of 9.881 million, reflecting an urbanization rate of 72.2%. The distribution across primary, secondary, and tertiary industries is delineated as 1.5:

Y. Bo et al., *Pathways for Reaching Air Quality Standard and Carbon Peak in Selected Cities in China*, Urban Sustainability, https://doi.org/10.1007/978-981-96-8315-4_3

43.9: 54.7, with coal consumption constituting 63.2% of the overall energy consumption. Notably, the vehicular count surpasses four million, securing the city's position as the seventh-ranking locale nationwide.

The annual pollution indicators for Zhengzhou City, covering the period from 2013 to 2017, are presented in Table 3.1. It is discernible that the yearly indicators for SO_2 and CO in Zhengzhou City have already achieved compliance with Grade II national standards for air quality. However, despite these achievements, observable breaches in air quality persisted in Zhengzhou City until the close of 2017. Specifically, in 2017, the annual mean concentration of $PM_{2.5}$ stood at 66 μg/m^3, surpassing the limit of the Grade II national standards for air quality by 88.6%. The annual mean concentration of PM_{10} reached 118 μg/m^3, exceeding the corresponding limit by 68.6%. Furthermore, the annual mean concentration of NO_2 measured 54 μg/m^3, exceeding the prescribed limit by 35.0%. The 90th percentile of the maximum daily 8-h O_3 concentration reached 199 μg/m^3, exceeding the limit of the Grade II national standards for air quality by 24.3%.

In light of the extant air quality scenario and in alignment with viable strategies in Zhengzhou, coupled with pertinent initiatives such as the "13th Five-Year Ecological and Environmental Protection Plan of Henan Province", the "13th Five-Year Energy Development Plan of Zhengzhou", the "Three-Year Action Plan for Pollution Prevention and Control in Henan Province (2018–2020)", and the "Three-Year Action Plan for Winning the Blue Sky Defense Battle in Zhengzhou (2018–2020)", Zhengzhou has devised air quality compliance objectives. By the year 2028, concentrations of $PM_{2.5}$ and PM_{10} are projected to essentially meet the Grade II national standards for air quality, while concentrations of SO_2, CO, and NO_2 are expected to consistently conform to the requirements of the Grade II national standards for air quality. Looking ahead to 2035, concentrations of O_3 are anticipated to satisfy the stipulations of the Grade II national standards for air quality.

Table 3.1 Statistics of annual pollution indicators in Zhengzhou City from 2013 to 2017

Pollutants	2013	2014	2015	2016	2017
PM2.5	108	88	96	78	66
PM_{10}	171	158	167	143	118
O_3_8h_90per	109	116	159	177	199
NO_2	52	51	58	56	54
SO_2	59	43	33	29	21
CO_95per	4.9	3.1	2.7	2.8	2.2

Notes:
1. The parameter O_3_8h_90per signifies the 90th percentile of the maximum daily 8-hour O_3 concentration, while CO_95per denotes the 95th percentile of daily average CO concentration
2. Concentrations of $PM_{2.5}$, PM_{10}, O_3_8h_90per, NO_2, and SO_2 are expressed in μg/m^3, while the concentration of CO_95per is expressed in mg/m^3

3.2 Analysis of the Emission Characteristics for the Baseline Year in Zhengzhou City

In Fig. 3.1, the contribution rates of atmospheric pollutants and CO_2 emissions in Zhengzhou City for the year 2017 are graphically represented. Notably, process sources and stationary combustion sources emerge as the predominant contributors to SO_2 emissions, contributing 50.0% and 40.9% to the overall emissions, respectively. Mobile sources assume a preeminent role in NO_x emissions, representing a substantial 57.6% of the total NO_x emissions in Zhengzhou. Furthermore, for CO and VOCs, process sources dominate as the primary emitters, accounting for 42.5% and 58.6% of their total emissions, respectively. Agricultural sources assume prominence as the principal contributors to NH_3 emissions in Zhengzhou, comprising a substantial 78.2% of the total emissions. Dust sources emerge as the predominant contributors to PM_{10} and $PM_{2.5}$, constituting up to 60.0% and 48.9% of the respective total emissions of these two pollutants. Mobile sources again dominate, contributing approximately 69.0% of the total black carbon emissions. Cooking oil fumes emanating from catering sources stand out as the primary contributors to organic carbon emissions, comprising 50.1% of the total emissions. In the realm of CO_2 emissions, stationary combustion sources take precedence, constituting a noteworthy 73.26% of the total emissions in Zhengzhou.

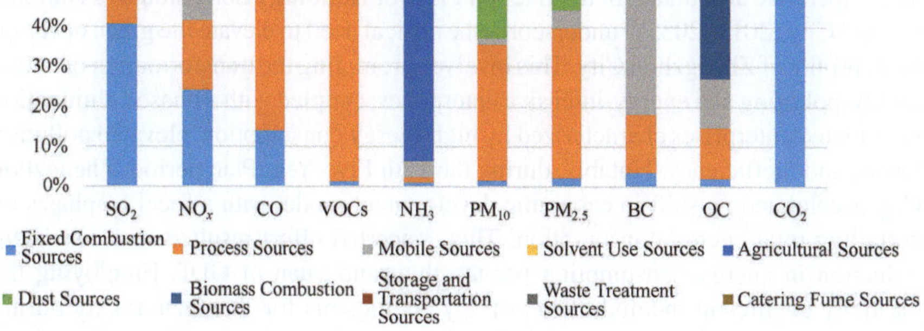

Fig. 3.1 The contribution rates of atmospheric pollutants and CO_2 emissions in Zhengzhou City in 2017

3.3 Forecast of Economic and Energy Development

3.3.1 Forecast of Economic Development

Aligned with the imperatives articulated in the "Action Plan for Building Zhengzhou as a National Central City (2017–2035)", Zhengzhou City embarks upon a transformative trajectory to evolve into a national central city comprehensively. The envisioned trajectory spans the period from 2020 to 2035, during which Zhengzhou aspires to ascend to the status of a national central city, positioning itself amidst the vanguard of national innovative cities. Concurrently, Zhengzhou seeks to establish itself as an international comprehensive hub, a pivotal international logistics center, a substantive national economic growth hub, and a highly dynamic national innovation and entrepreneurship center.

Guided by population policies, including the comprehensive two-child policy, and synergizing with empirical population growth data, projections indicate that the population of Zhengzhou City will approach 12.614 million by the year 2028. Within this estimate, the urban populace is anticipated to reach approximately 10.432 million, thereby achieving an urbanization rate of 82.7%. Factoring in annual growth rates derived from China's GDP for the delineated periods of 2016–2020 (6.8%), 2021–2025 (5.5%), and 2026–2028 (4.5%), alongside historical annual GDP growth rates in Zhengzhou City, it is forecasted that the GDP of Zhengzhou City will ascend to approximately 2232.98 billion yuan in the year 2028.

3.3.2 Forecast of Energy Development

The imperative articulated in the "Action Plan for Building Zhengzhou as a National Central City (2017–2035)" underscores the critical need to elevate the green development profile of Zhengzhou City. This involves promoting the transformation of extant highly-polluting and energy-intensive enterprises, coupled with a phased elimination of outdated enterprises characterized by high energy consumption, elevated pollution levels, and inefficiency. Notably, during the 12th Five-Year Plan period, Zhengzhou City accelerated its shift in economic development mode, with a focal emphasis on curtailing total energy consumption. This concerted effort resulted in a consistent reduction in energy consumption per ten thousand yuan of GDP. Employing the elasticity coefficient method, the elasticity coefficients for Zhengzhou City during the 12th Five-Year Plan period, the 13th Five-Year Plan period, and the subsequent period from 2021 to 2028 are calculated as 0.33, 0.31, and 0.21, respectively. A forecasted projection places the total energy consumption of Zhengzhou City at 40.536 million tons of standard coal in the year 2028.

In 2017, the total energy consumption of Zhengzhou City amounted to 24.75 million tons of standard coal, with coal consumption constituting 63.2%, oil consumption at 14.3%, natural gas consumption at 11.7%, and primary electricity

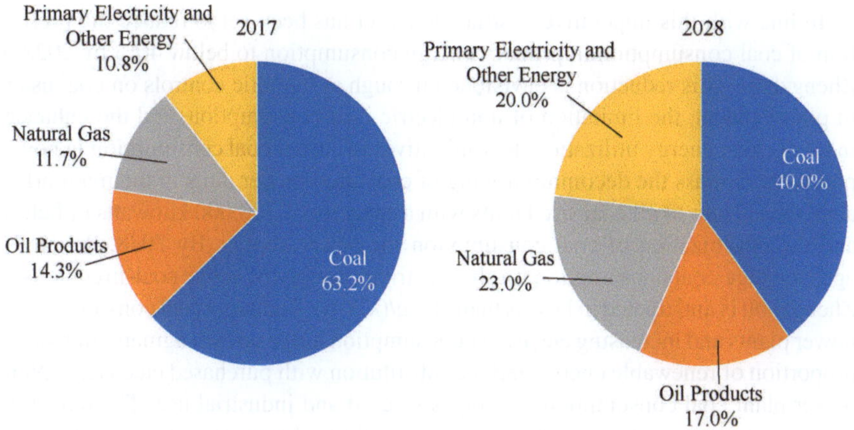

Fig. 3.2 The proportions of various energy consumption categories in Zhengzhou City for the years 2017 and 2028

and other energy consumption collectively representing 10.8% of the total energy consumption. Building upon this energy structure and informed by governmental research, the anticipated energy composition for Zhengzhou City in 2028 envisages coal consumption at 40.0%, oil consumption at 17.0%, natural gas consumption at 23.0%, and primary electricity and other energy consumption at 20.0%. The comparative proportions of various energy consumption categories in Zhengzhou City for the years 2017 and 2028 are visually depicted in Fig. 3.2.

3.4 Key Emission Reduction Tasks and Measures

3.4.1 Facilitating the Swift Adjustment of the Energy Structure and the Establishment of a Clean and Low-Carbon Energy System

As of 2017, coal consumption in Zhengzhou constituted a substantial 63.2% of the total energy consumption, with natural gas accounting for 11.7%. The overall proportion of clean energy consumption was comparatively low. The coal consumption in Zhengzhou is predominantly sourced from industries such as electricity and heat production, non-ferrous metal smelting, non-metallic mineral manufacturing, and coal mining and washing. Specifically, coal consumption in Zhengzhou's electricity sector represented 70% of the total raw coal consumption, notably trailing behind global benchmarks (94% in the U.S. and 81% in the EU). This underscores the imperative for optimizing the energy structure and coal combustion profile, necessitating a concerted effort to promote the efficient and clean utilization of coal resources.

In line with this imperative, a strategic target has been set to reduce the proportion of coal consumption in primary energy consumption to below 40% by 2028 in Zhengzhou. This reduction is envisioned through systematic controls on coal usage in power plants, the limitation of non-electric coal consumption, and the enhancement of clean energy utilization. Key initiatives to curtail coal consumption in power plants encompass the decommissioning of coal-fired power units in the main urban area, the closure of all coal-fired units with a capacity of 300,000 kilowatts or below, and the optimization of coal consumption for power supply. By 2028, the envisaged average coal consumption for the electricity supply of active coal-fired units in Zhengzhou is anticipated to be less than 280 g/(kW·h). Reducing coal consumption in power plants and increasing electricity consumption hinge on the augmentation of the proportion of renewable energy and the substitution with purchased electricity. Non-power plant coal consumption encompasses civil and industrial use. The reduction of coal consumption for civil heating will be achieved through robust development in combined heat and power centralized heating projects, advancement in renewable energy heating initiatives, and the steadfast implementation of clean energy heating in rural areas. By 2028, Zhengzhou aims to attain a 100% clean heating rate in rural areas. Efforts to diminish coal consumption in industrial enterprises center on the substitution of coal with clean energy and the improvement of energy efficiency. By 2028, Zhengzhou aspires to realize a 35% reduction in energy consumption per unit of GDP compared to the 2017 baseline. Moreover, the city needs to advance energy-saving and low-carbon buildings. This encompasses not only the execution of energy-saving retrofits for existing structures but also the expansion of green buildings and the initiation of pilot projects for ultra-low energy consumption or nearly zero-energy buildings, along with the incorporation of passive solar house designs.

3.4.2 Realigning and Enhancing the Industrial Framework for a Green and Low-Carbon Industrial System

The existing industrial configuration in Zhengzhou City is characterized by an inherent irrationality, exemplified by the prevalence of an industrial siege phenomenon, where industries with high pollution and energy consumption feature prominently. In light of this, it is imperative for Zhengzhou to undertake a comprehensive optimization of its industrial layout and a strategic adjustment of the industrial structure.

Regarding the optimization of the industrial layout, Zhengzhou City is conceptually divided into four distinct areas: the central urban area, the western region, the eastern region, and the southern region. The central urban area encompasses the five districts within Zhengzhou City and the Zhengdong New District, with a primary focus on establishing the core functional zone of the urban area while phasing out all inefficient production capacity. The western region, comprising the High-tech Zone, Shangjie District, Xingyang City, Dengfeng City, Xinmi City, and

Gongyi City, is strategically oriented towards nurturing industrial clusters at the trillion-yuan and hundred-billion-yuan levels, along with the development of several ten-billion-yuan industrial parks. The eastern region, consisting of the Economic Development Zone and Zhongmou County, aims to relieve certain inefficient functions while propelling the expansion and development of the peripheral International Logistics Park and Zhongmou Automobile Industry Cluster. The southern region, encompassing the Airport Economy Zone and Xinzheng City, is envisioned as a future emerging industry agglomeration and expansion zone for Zhengzhou City. A key objective by 2028 is to transition significant industry enterprises in each county (city, district) of Zhengzhou into industrial parks based on their leading functions.

In terms of adjusting the industrial structure, stringent environmental access standards are enforced at the source, accompanied by a strict prohibition on the entry of highly polluting and energy-intensive enterprises. This is achieved through enhanced constraints and guidance of the atmospheric environmental quality on planning and environmental assessments, rigorous control of capacity in high-pollution and high-energy-consumption industries, strict regulations on coal-fired projects, and a tightening of the construction of projects associated with VOCs. Moreover, concerted efforts are directed towards controlling outdated, inefficient, and excess production capacity. Specifically, excess capacity in industries such as cement, brown fused alumina, lime, stone, chlorinated paraffin, casting, abrasive tools, electrolytic aluminum, carbon, refractory materials, bricks, and tiles in Zhengzhou is targeted for elimination, thereby raising the overall industry standards. Concurrently, measures are implemented to forestall the resurgence of "scattered, non-compliant, and polluting" enterprises, and proactive development of energy-saving, environmentally friendly, and green, low-carbon industries is actively pursued.

3.4.3 Enhancing Pollution Mitigation in Key Sectors and Thoroughly Executing the Rectification of Volatile Organic Compounds

Zhengzhou City boasts a substantial industrial scale, with emissions from pivotal monitored enterprises contributing a noteworthy one-tenth to the total emissions of Henan Province. In the year 2017, the collective emissions of SO_2, NO_x, particulate matter, smoke, and dust in Zhengzhou ranked second in Henan Province. This cumulative pollutant release surpassed the atmospheric environmental capacity aligned with the socioeconomic development of Zhengzhou City. Consequently, alongside the imperative of adjusting the industrial structure, there is a pressing need for Zhengzhou City to elevate its commitment to clean production. This entails substantial measures such as the upgrading and overhaul of non-electric industries, including but not limited to steel, cement, non-ferrous metals, refractories, and ceramics. Simultaneously, dedicated initiatives are directed towards remediating industrial furnaces and kilns, complemented by a focused campaign aimed at controlling volatile organic compounds. By the year 2028, it is anticipated that key industry enterprises in Zhengzhou City will attain ultra-low emissions, encompassing both organized and unorganized operations, and spanning the entirety of material transportation systems.

3.4.4 Proactively Restructuring the Transportation Framework and Advancing the Green and Low-Carbon Transportation System

As of 2018, the predominant share of road freight volume in Zhengzhou City stood at 92.9%, followed by railway freight volume at 7.0%, and civil aviation freight volume at 0.1%. This underscores a prevailing reliance on road transport for goods movement, emphasizing the imperative for strategic adjustments in the transportation structure. Key initiatives to address this concern involve a substantial increase in the proportion of railway freight, the robust development of multimodal transportation, and the establishment of an environmentally sustainable urban logistics system. By the year 2028, the anticipated proportion of railway freight volume is set to reach 35%, with large-scale major production enterprises in sectors such as steel, power, and electrolytic aluminum contributing over 50% of the railway freight volume. Simultaneously, backed by railway logistics bases and highway ports, the construction of multimodal and seamlessly interconnected freight hubs (logistics parks) is advanced. Pilot programs are initiated for integrated rail and road transportation of urban production and daily necessities, marking the establishment of a new urban distribution system characterized by road-rail intermodal transport featuring "external collection, internal distribution, and green intermodal transportation".

Motor vehicle ownership in Zhengzhou City has surpassed four million, with the contribution of motor vehicle emissions to $PM_{2.5}$ reaching 24.5% based on the results of the $PM_{2.5}$ emission source analysis in Zhengzhou, necessitating a comprehensive approach to control motor vehicle quantity. This primarily involves fortifying the green transportation system and promoting the adoption of new energy vehicles. By 2028, it is anticipated that the motorization share of public transport in Zhengzhou will exceed 73%, with plans being explored to designate "zero-emission driving zones" in concentrated office or commercial areas. Concurrently, the strategic plan and implementation scheme are imperative for controlling the number of fuel-powered motor vehicles in Zhengzhou City, with a targeted annual increase in the proportion of new energy passenger car acquisitions. As of 2020, all public buses, taxis, municipal sanitation vehicles, and light-duty logistics delivery vehicles in the urban built-up area of Zhengzhou City have transitioned to full electrification. The proportion of new energy passenger car acquisitions in Zhengzhou is expected to surpass 80% by 2028 and reach 100% by 2035. Given the substantial contribution of old vehicles and high-emission diesel vehicles to atmospheric pollutant emissions, concerted efforts are dedicated to phasing out outdated vehicles and addressing heavy-duty diesel vehicles. By 2025, Zhengzhou aims to retire all operational diesel vehicles compliant with China III vehicle emission standards and below, while promoting the retrofitting of high-emission vehicles with diesel particulate filters and selective catalytic reduction devices. These measures are anticipated to yield significant reductions in NO_x and particulate matter emissions from the vehicle fleet.

3.4.5 Optimizing and Restructuring Land Use Allocation while Advancing the Control of Non-point Source Pollution

Zhengzhou City is characterized by a substantial number of construction sites, many of which are linear projects and contiguous construction sites dispersed across various urban zones. Despite initial efforts to standardize construction site management, the persistent challenge of dust pollution from construction activities persists. Moreover, incomplete coverage of construction and debris hauling vehicles has led to instances of dust dispersion, posing challenges in the effective management of road dust. In response, the focus of dust management in Zhengzhou City revolves around reinforcing construction dust control, refining road dust management, and implementing dust control measures for storage yards. This entails the comprehensive implementation of the "Eight 100%" initiative, basically achieving mechanized cleaning of road surfaces, reducing the density of floating dust on road surfaces after mechanized cleaning to below 3 g/m^2, and ensuring 100% coverage and networking for online monitoring and video surveillance of coal yards, material yards, and slag yards in key areas.

To mitigate agricultural non-point source pollution, the management measures are threefold. The first involves the prohibition of open burning of crop straws and the establishment of an improved system for collecting and storing straws, with an emphasis on promoting resource utilization. The second measure entails strengthening the resource utilization of livestock and poultry manure, with an anticipated achievement of ecological consumption or emission compliance of livestock and poultry manure in Zhengzhou City by 2028. The third measure advocates for the reduction and improved efficiency of chemical fertilizers and pesticides in crop cultivation, aspiring to achieve negative growth in their use. By 2028, the fertilizer utilization rate in Zhengzhou City is projected to surpass 50%.

In tandem with the adjustment of land use structure, there is an imperative for the implementation of green carbon sink projects. Expansion of urban green space involves meticulous planning for green spaces, demolishing illegal constructions to create green areas, integrating greenery wherever feasible, and allocating designated spaces for increased green coverage. The envisioned outcome by 2028 is a green space ratio (inclusive of three-dimensional greening and roof greening) in the urban built-up area of Zhengzhou City reaching 40.0%.

3.5 Analysis of Emission Reduction Potential

Drawing from the atmospheric pollutant emission inventory of Zhengzhou City in 2017 and projections for incremental emissions from motor vehicles and natural gas in 2028, coupled with the outlined adjustment plans for energy, industrial, transportation, and land use structures, along with the ongoing status of end-of-pipe treatment in key industries, estimations reveal that by 2028, Zhengzhou is poised to achieve substantial reductions of 74.5%, 70.4%, 56.4%, 43.2%, and 65.7% in SO_2, NO_x, VOCs, NH_3, and primary $PM_{2.5}$ emissions, respectively, relative to the baseline year.

A detailed examination of diverse management measures underscores the pivotal role of industrial upgrading and transformation as the most influential factor in reducing SO_2 emissions, contributing to approximately 38.5% of the overall emissions reduction. For NO_x emissions reduction, the primary measure revolves around controlling pollution from mobile sources, constituting approximately 52.9% of the total emissions reduction. In the realm of VOCs emissions reduction, the treatment of volatile organic compounds emerges as the most impactful measure, representing around 33.5% of the total emissions reduction. NH_3 emissions reduction is predominantly influenced by the control of agricultural non-point source pollution, contributing approximately 86.5% to the overall emissions reduction. Regarding $PM_{2.5}$ emissions reduction, comprehensive dust control stands out as the predominant measure, making up approximately 41.1% of the total emissions reduction. In the case of CO_2 emissions reduction, the most substantial contribution emanates from the adjustment and comprehensive treatment of the power structure, comprising approximately 55.4% of the total emissions reduction (see Fig. 3.3).

3.6 Analysis of Air Quality Compliance

Employing the iterative calculation method for the atmospheric environmental capacity within the WRF-CMAQ air quality model, coupled with simulations of the average proportions of sulfate, nitrate, ammonium, and primary $PM_{2.5}$ to $PM_{2.5}$ in Zhengzhou, a strategic approach is devised to achieve compliance with the annual average concentration standard for $PM_{2.5}$ in Zhengzhou. This involves the formulation of emission reduction plans for SO_2, NO_x, NH_3, primary $PM_{2.5}$, and VOCs. Based on spatially differentiated emission reduction plans, new emission inventories are iteratively created and subsequent numerical simulation iterative calculations are carried out.

The results of the iterative simulations reveal that by reducing SO_2, NO_x, NH_3, primary $PM_{2.5}$, and VOCs by 60.0%, 70.0%, 40.0%, 65.0%, and 50.0%, respectively, relative to the baseline emissions in 2017, the simulated $PM_{2.5}$ concentration falls below the compliance limit of 35.0 $\mu g/m^3$. Based on the evaluation of emission reduction measures, it is projected that atmospheric pollutant emissions in Zhengzhou City

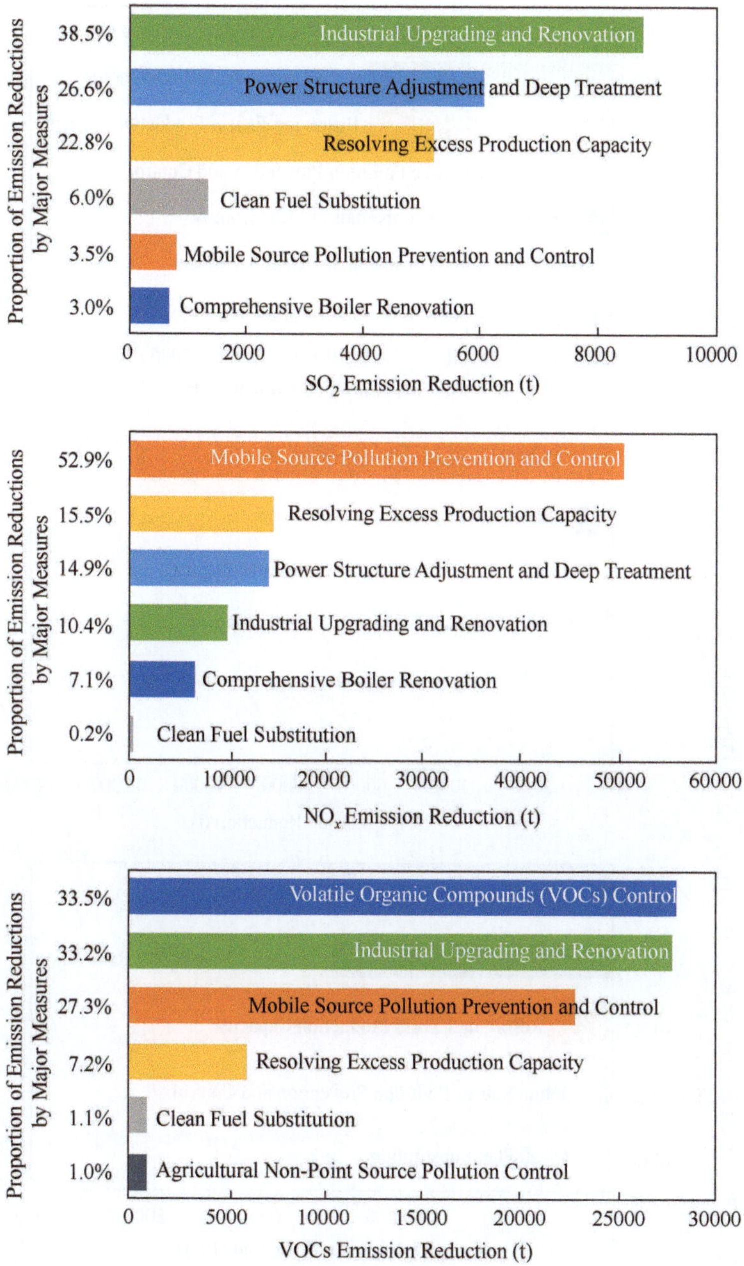

Fig. 3.3 Evaluation of the impact of key measures on reduction of atmospheric pollutants and CO_2 emissions

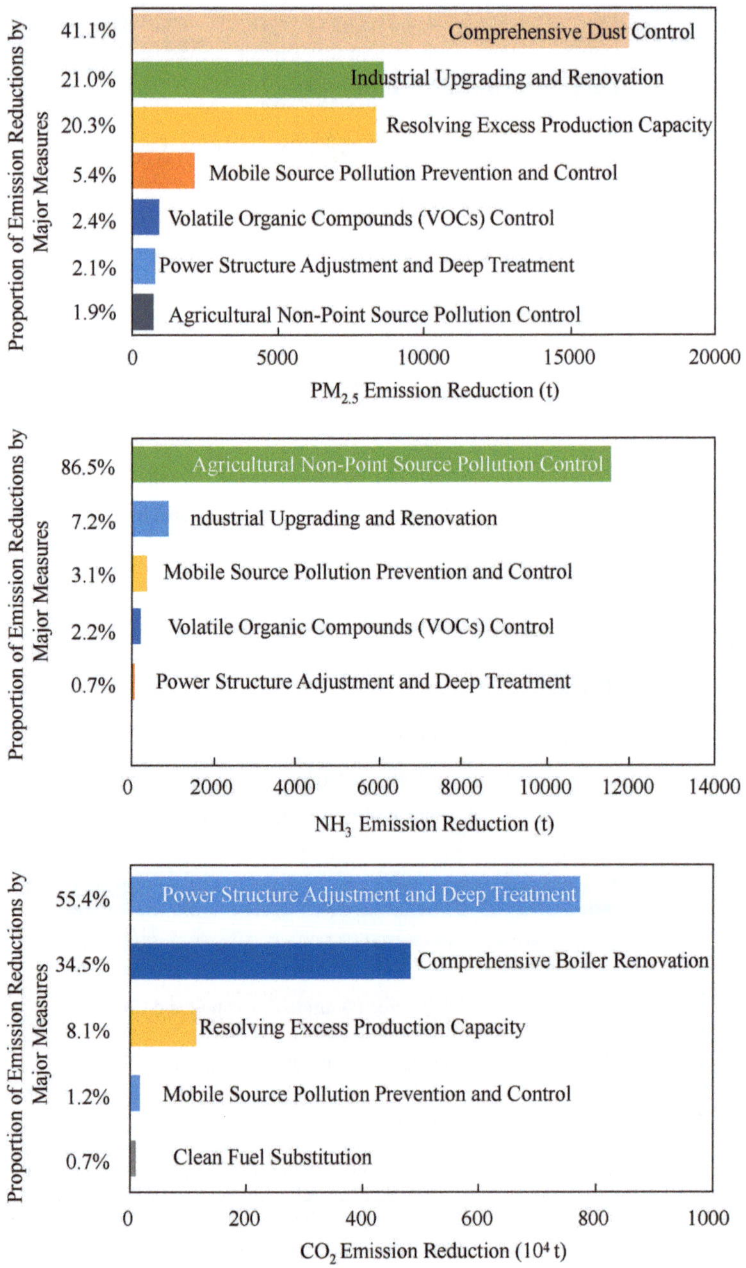

Fig. 3.3 (continued)

in 2028 will align with the atmospheric environmental capacity under the compliance state.

Drawing on the aforementioned ratios of pollutant emission reduction, this investigation formulates an emission inventory for Zhengzhou in 2028 and undertakes simulations utilizing an air quality model. Simultaneously, an assumption is made that the concentration of atmospheric pollutants in Zhengzhou City, influenced by the transport of surrounding pollution, will diminish by 45% in the year 2028. Through a comprehensive consideration of both localized emission reductions in Zhengzhou City and reductions in the adjacent regions, an analysis of the annual average simulated concentrations of major atmospheric pollutants in Zhengzhou for the year 2028 is executed.

As per the simulation results, following the implementation of pollution control measures, the $PM_{2.5}$ concentration is projected to experience a substantial reduction of 49.5% relative to the levels observed in 2017, reaching approximately 33.3 μg/m^3 by the year 2028. A closer examination, focusing exclusively on local emission reductions, reveals an anticipated decrease in $PM_{2.5}$ concentration of 21.9 μg/m^3, with an estimated concentration of 44.1 μg/m^3. Furthermore, the reduction in $PM_{2.5}$ concentration influenced by diminished surrounding pollution transport is expected to be 10.8 μg/m^3. Concerning other pollutants, the annual average simulated concentration of PM_{10} is estimated to decrease by 46.7% compared to 2017, with a projected concentration of 62.9 μg/m^3 in 2028. Similarly, the annual average simulated concentration of NO_2 is forecasted to decrease by 45.8%, reaching an estimated concentration of 29.3 μg/m^3 in 2028. The annual average simulated concentration of SO_2 is expected to undergo a considerable decline of 66.6%, with an estimated concentration of 7.0 μg/m^3 in 2028. The 95th percentile of daily CO concentration is projected to decrease by 45.8%, with an estimated concentration of 1.2 mg/m^3 in 2028. Additionally, the 90th percentile of the maximum daily 8-hour O_3 concentration is expected to decrease by 15.0%, with an estimated concentration of 166.9 μg/m^3 in 2028.

In conclusion, the major air pollutants in Zhengzhou, encompassing $PM_{2.5}$, PM_{10}, SO_2, NO_2, and CO, are anticipated to meet the national ambient air quality standards by the year 2028, with effective control of O_3 pollution. In tandem with consolidating current achievements in air pollution control, Zhengzhou City is actively intensifying the adjustment of its energy structure and optimizing industrial and transportation systems. This includes the phased removal of old vehicles, the promotion of new energy vehicles, and the complete substitution with low-VOC raw and auxiliary materials. Additionally, Zhengzhou City is committed to fostering collaborative efforts for the prevention and control of regional air pollution, innovating environmental management policies and measures, and enhancing the proactive pollution control initiatives of enterprises. Through the implementation of these measures, the O_3 concentration in Zhengzhou is anticipated to achieve the Grade II national standards for air quality by the year 2035.

3.7 Analysis of Predictions for Peak Carbon Emissions

The LEAP model is employed to simulate and project long-term carbon emissions scenarios for Zhengzhou City (Fig. 3.4), encompassing indirect carbon emissions from purchased electricity in the emission inventory. In the BAU scenario, characterized by minimal improvements in the energy structure and energy intensity across various sectors, carbon emissions are anticipated to continue increasing rapidly in tandem with economic development and rising energy demand. Conversely, in the AQP scenario, where Zhengzhou achieves compliance with air quality standards by 2028, carbon emissions in the city are expected to peak in 2025 at 79.7 million tons, followed by a gradual decline. Relative to the simulated results under the BAU scenario, carbon emissions in Zhengzhou City experience reductions of 3.32, 9.78, and 17.01 million tons in 2020, 2025, and 2030 under the AQP scenario, respectively.

In the AQP scenario, the carbon emissions by sector in Zhengzhou City are depicted in Fig. 3.5.

The carbon emissions from the industrial sector have witnessed a substantial decrease since 2017, leading to a reduction of 14.12 million tons in carbon emissions by 2028 compared to the baseline scenario. This reduction in carbon emissions from the industrial sector is attributed to two main factors. Firstly, there is an overall decrease in energy demand for terminal industrial use, primarily stemming from the elimination of inefficient, surplus, and outdated capacities in high-energy-consuming industries like cement and non-ferrous metals, along with a decline in energy consumption per unit of product or value added. The second factor involves the transition to a cleaner power generation structure. By 2035, the proportion of coal-fired power generation in Zhengzhou City is expected to decrease to 50%, resulting in substantial benefits in terms of carbon reduction.

In the civil sector and public building sector, air quality planning primarily concentrates on curbing pollutant emissions within the construction sector through the endorsement of electrification, encompassing initiatives like coal-to-gas and coal-to-electricity. Concurrently, with considerations for energy-efficient upgrades in buildings and advancements in energy-saving technologies for household appliances,

Fig. 3.4 Long-term carbon emissions scenarios for Zhengzhou City simulated and predicted using the LEAP model

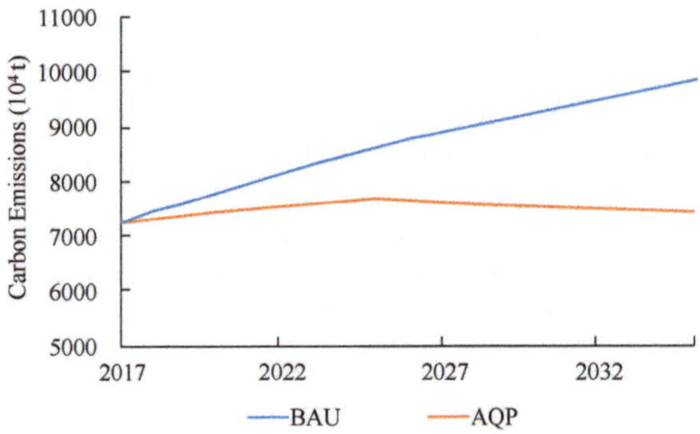

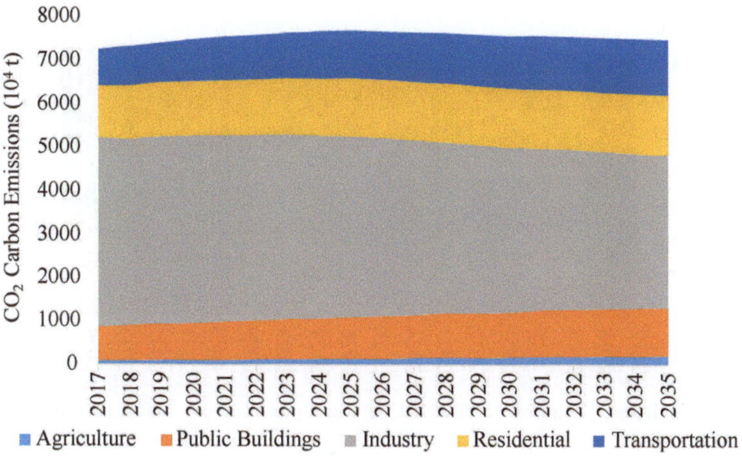

Fig. 3.5 Carbon emissions by sector in Zhengzhou City under the AQP scenario

the construction sector is anticipated to contribute to a cumulative carbon emission reduction of 1.17 million tons relative to the baseline scenario. However, in light of advancing living standards and the rapid growth of the tertiary industry, the overall energy consumption demand within the construction sector is expected to persist in its upward trajectory, resulting in a delay in the peaking of carbon emissions, likely extending beyond 2035.

Regarding the transportation sector, Zhengzhou intends to revamp its transportation structure by promoting the widespread adoption of new energy vehicles in both freight and public transportation. This initiative is anticipated to yield a carbon emission reduction of 72,000 tons by 2028 in comparison to the baseline scenario. However, despite these efforts, the overall energy demand and carbon emissions from the transportation sector in Zhengzhou are expected to rise due to the projected increase in the number of motor vehicles.

In the AQP scenario, Fig. 3.6 illustrates the carbon emissions from various energy sources and industrial processes in Zhengzhou City. Carbon emissions from industrial processes are predominantly linked to cement production. Generally, carbon emissions from industrial processes and coal consumption exhibit a consistent year-over-year decline, while emissions from oil, natural gas, and purchased electricity show an increasing trend annually. By 2035, the total purchased electricity in Zhengzhou City is projected to increase from 15 billion kW·h in 2017 to approximately 25 billion kW·h. Although this study does not account for changes in the emission factors of the power grid due to the limited impact of individual city policies on the overall emission factors of the power grid, the anticipated shift toward a cleaner national power structure is expected to decrease the carbon emission factors of regional power grids. Consequently, the indirect carbon emissions from purchased electricity in Zhengzhou City are projected to be lower than the current estimates.

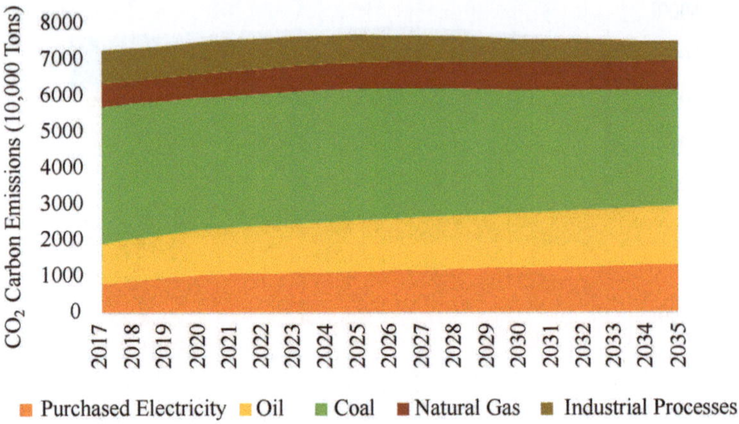

Fig. 3.6 Carbon emissions from diverse energy sources and industrial processes in Zhengzhou City under the AQP scenario

Chapter 4
Analysis of the Pathways to Achieving "Dual Attainments" in Shijiazhuang City

Abstract This chapter delves into the paradigm of "dual attainments" within the context of heavy industrial cities, with a specific focus on Shijiazhuang City. Through an analysis encompassing the general landscape of Shijiazhuang City, the present condition of environmental air quality, the distinctive features of atmospheric pollutant emissions, and the established timelines for meeting air quality standards, strategic approaches are proposed to facilitate Shijiazhuang City's achievement of air quality compliance. These strategies are formulated based on comprehensive assessments derived from urban emission inventories and model simulations. Additionally, the chapter explores the prospective scenario of achieving peak carbon emissions while ensuring compliance with air quality standards.

According to model calculations, achieving air quality compliance in Shijiazhuang City necessitates substantial reductions in the emissions of various pollutants relative to their 2017 baseline levels. Specifically, reductions of 60%, 70%, 45%, 70%, and 50% respectively for SO_2, NOx, NH_3, primary $PM_{2.5}$, and VOCs are required. Upon attaining these reductions, simulated PM2.5 concentrations are anticipated to fall below 35.0 μg/m^3. Projections indicate that by 2033, the annual average concentrations of $PM_{2.5}$, PM_{10}, SO_2, CO, and NO_2 will align with regulatory standards. These concentrations are anticipated to reach 31 μg/m^3, 58 μg/m^3, 6.5 μg/m^3, 0.52 mg/m^3, and 23.2 μg/m^3, respectively. However, it is noted that the 90th percentile of the maximum daily 8-hour O_3 concentration is expected to rise compared to the 2017 baseline level. Coordinated measures aimed at reducing VOCs emissions are recommended to bring the O_3 level into compliance by 2035. Regarding peak carbon emissions, with the target of achieving air quality compliance by 2033, it is forecasted that emissions in Shijiazhuang City will reach their maximum in 2020, followed by a sustained decline.

Y. Bo et al., *Pathways for Reaching Air Quality Standard and Carbon Peak in Selected Cities in China*, Urban Sustainability, https://doi.org/10.1007/978-981-96-8315-4_4

4.1 Overview of Shijiazhuang City

Shijiazhuang City, situated in the southern-central region of Hebei Province, serves as the capital of the province. Geographically, it shares boundaries with Hengshui City to the east, connects with Xingtai City to the south, borders Shanxi Province to the west, and has a northern border with Baoding City. Located at a distance of 283 kilometers from Beijing, the city encompasses a total area of 15,848 square kilometers. As of the conclusion of 2017, Shijiazhuang City boasted a permanent resident population of 10.88 million, accompanied by an urbanization rate of 61.6% among the permanent residents. In 2017, the sectoral distribution of economic activities exhibited a composition of 7.4% in the primary industry, 45.1% in the secondary industry, and 47.5% in the tertiary industry. Notably, enterprises above the designated size recorded an energy consumption of 23.978 million tons of standard coal, with the six major high-energy-consuming industries contributing significantly, comprising 86.0% of the overall energy consumption. Furthermore, the city's vehicular landscape witnessed the registration of 2.56 million vehicles, securing the 13th position nationwide.

Table 4.1 delineates the annual pollution indicators for various pollutants in Shijiazhuang City spanning the years 2013 to 2017. Notably, the annual average indicators for SO_2 and CO in Shijiazhuang City have already achieved compliance with Grade II national standards for air quality. Despite this accomplishment, the termination of 2017 witnessed discernible breaches in air quality within Shijiazhuang City. In particular, during 2017, the annual average concentration of $PM_{2.5}$ in Shijiazhuang City escalated to 86 $\mu g/m^3$, exceeding the limit of the Grade II national standards for air quality by a notable margin of 145.7%. Simultaneously, the annual average concentration of PM_{10} reached 158 $\mu g/m^3$, surpassing the stipulated limit by 125.7%. The annual average concentration of NO_2 was recorded at 54 $\mu g/m^3$, exceeding the established limit by 35.0%. Furthermore, the 90th percentile of the maximum daily 8-hour O_3 concentration soared to 201 $\mu g/m^3$, surpassing the limit of the Grade II national standards for air quality by 24.3%.

Given the existing air quality scenario in Shijiazhuang City, coupled with viable measures, and aligned with pertinent initiatives such as the "13th Five-Year Ecological and Environmental Protection Plan of Hebei Province", the "13th Five-Year Energy Development Plan of Hebei Province", the "Three-Year Action Plan for Winning the Blue Sky Defense Battle in Hebei Province", and the "Three-Year Action Plan for Winning the Blue Sky Defense Battle in Shijiazhuang City (2018–2020)", Shijiazhuang City has delineated comprehensive air quality compliance objectives. Specifically, by the year 2033, the city aims to achieve annual average concentrations of $PM_{2.5}$ and PM_{10} that essentially meet the Grade II national standards for ambient air quality. Additionally, the anticipated trajectory includes ensuring that the annual average concentrations of SO_2, CO, and NO_2 consistently adhere to the Grade II national standards for air quality. By the year 2035, it is projected that O_3 concentrations will align with the Grade II national standards for ambient air quality.

Table 4.1 Statistics of annual pollution Indicators for various pollutants in Shijiazhuang City from 2013 to 2017

Pollutants	2013	2014	2015	2016	2017
PM2.5	154	124	89	99	86
PM_{10}	305	206	147	164	158
SO_2	105	62	47	41	33
NO_2	68	53	51	58	54
O_3_8h_90per	173	161	148	164	201
CO_95per	5.7	4.2	4.3	3.9	3.6

Notes:
1. The parameter O_3_8h_90per signifies the 90th percentile of the maximum daily 8-hour O_3 concentration, while CO_95per denotes the 95th percentile of daily average CO concentration
2. Concentrations of $PM_{2.5}$, PM_{10}, O_3_8h_90per, NO_2, and SO_2 are expressed in μg/m^3, while the concentration of CO_95per is expressed in mg/m^3

4.2 Analysis of the Emission Characteristics for the Baseline Year in Shijiazhuang City

The contribution rates of atmospheric pollutants and CO_2 emissions in Shijiazhuang City during the year 2017 are visually represented in Fig. 4.1. Notably, stationary fossil fuel combustion emerges as the predominant source of SO_2 emissions, primarily emanating from civil combustion, industrial boilers, and electric heating, with a substantial contribution rate of 64.7%. Mobile sources, encompassing heavy-duty freight vehicles, light-duty freight vehicles, and tractors, emerge as the primary contributors to NO_x emissions, collectively constituting 42.9% of the overall NO_x emissions. Dust emissions play a pivotal role in influencing PM_{10} levels in Shijiazhuang City, contributing significantly with a rate of 50.7%. Process sources stand out as the principal contributors to both $PM_{2.5}$ and VOCs emissions, accounting for 35.6% and 54.8%, respectively. Agricultural sources take precedence in NH_3 emissions, representing a dominant share of 86.0% of the total emissions. Furthermore, when considering CO_2 emissions, stationary fossil fuel combustion sources emerge as the major contributors in Shijiazhuang City, commanding a noteworthy contribution rate of 80.3%.

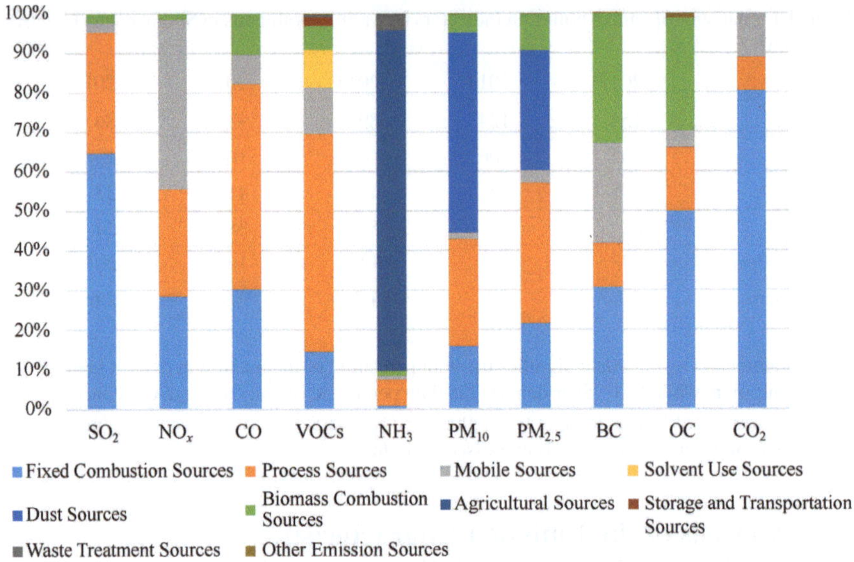

Fig. 4.1 The contribution rates of atmospheric pollutants and CO_2 emissions in Shijiazhuang City in 2017

4.3 Forecast of Economic and Energy Development

4.3.1 Forecast of Economic Development

As outlined in the "Overall Urban Planning of Shijiazhuang (2018–2035)" and informed by simulation outcomes from the China Population Projection System (CPPS), the population trajectory of Shijiazhuang City, considering solely natural growth, is anticipated to crest around 2028, reaching an approximate figure of 10.936 million individuals. Subsequently, by the year 2033, the population is projected to reach around 10.933 million people. When factoring in the influence of the economic growth rate, shifts in industrial structure, and the efficiency of economic development on mechanical population growth, it is envisaged that the mechanical population growth of Shijiazhuang City will culminate around 2030, reaching an estimated 12 million people. Furthermore, a comprehensive forecast, incorporating the trend extrapolation method, the United Nations method, and the Logistic curve regression method, anticipates that the urbanization rate of Shijiazhuang City will ascend to 77% by 2033. This analysis further discerns that the urban population of Shijiazhuang City is forecasted to comprise approximately 9.25 million people by 2033.

As articulated in the "China Energy Outlook 2030", the anticipated annual average GDP growth rates for China are projected to be 6.8%, 5.5%, and 4.5% for the periods of 2016–2020, 2021–2025, and 2026–2030, respectively. Complementing this national perspective, insights from the "Shijiazhuang Statistical Yearbook" regarding the annual average growth rate of Shijiazhuang City's GDP during the period of 2008–2017 contribute to the forecasted trajectory. Applying the elasticity coefficient method, it is forecasted that the GDP of Shijiazhuang City will attain 1211.2 billion yuan by the year 2033.

4.3.2 Forecast of Energy Development

During the 12th Five-Year Plan period, Shijiazhuang City diligently executed the energy conservation and emission reduction directives set forth by the Hebei Provincial Government. The city accelerated the transformation of its economic development paradigm with a pronounced emphasis on curbing total energy consumption. This concerted effort resulted in a consistent reduction in energy consumption per ten thousand yuan of GDP. The "Comprehensive Plan for Energy Conservation and Emission Reduction during the 13th Five-Year Plan Period in Shijiazhuang City" articulated ambitious goals, proposing an 18% decrease in energy consumption per ten thousand yuan of GDP compared to the 2015 baseline. Additionally, the plan aimed to cap total energy consumption at 41.46 million tons of standard coal by the year 2020. This strategic framework encompassed a spectrum of initiatives, including the promotion of transformation and upgrades in traditional industries, driving the optimization of the energy structure, enhancing energy utilization efficiency, and implementing measures targeting both total energy consumption and intensity. Shijiazhuang City is anticipated to achieve a sustained reduction in energy consumption per ten thousand yuan of GDP from 2021 to 2033.

In the year 2017, the aggregate energy consumption in Shijiazhuang City amounted to 44.449 million tons of standard coal, with coal contributing significantly at 30.67 million tons of standard coal. By obtaining data on energy consumption per unit of GDP for Shijiazhuang City spanning the period from 2005 to 2017 from the official website of the Hebei Provincial Bureau of Statistics, it is projected that the energy consumption per ten thousand yuan of GDP in Shijiazhuang City will be 0.42 tons of standard coal, and the total energy consumption is estimated to reach approximately 50.953 million tons of standard coal by the year 2033 by utilizing the energy consumption per unit of GDP method. In terms of the prospects for specific energy sources, it is envisaged that coal consumption in Shijiazhuang City will decrease to 22.93 million tons of standard coal by 2033. The proportional representation of various energy consumption categories in Shijiazhuang City for the years 2017 and 2033 is elucidated in Fig. 4.2.

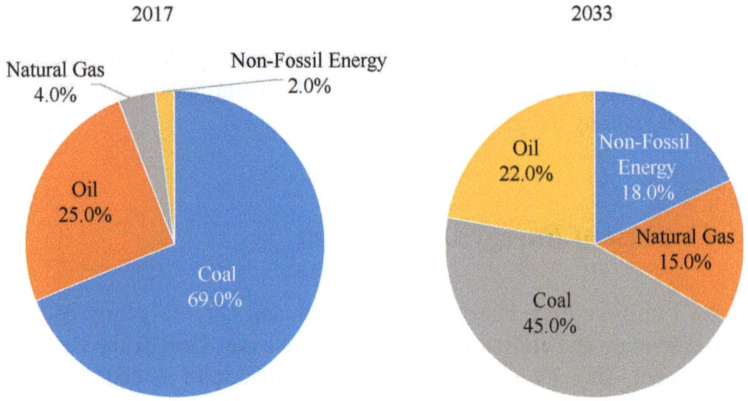

Fig. 4.2 The proportions of various energy consumption categories in Shijiazhuang City for the years 2017 and 2033

4.4 Key Emission Reduction Tasks and Measures

4.4.1 Facilitating the Swift Adjustment of the Energy Structure and the Establishment of a Clean and Low-Carbon Energy System

In 2017, coal consumption constituted a significant portion, accounting for 69.3% of the primary energy consumption in Shijiazhuang City. This elevated reliance on coal underscores the imperative for a prompt adjustment in the energy structure. The restructuring of the energy sector is principally achieved through a comprehensive approach, involving measures aimed at controlling total coal consumption, establishing a clean and low-carbon heating system, advocating for the adoption of renewable energy sources, and enhancing energy utilization efficiency.

Efforts to control total coal consumption encompass a range of strategies, including the regulation of coal usage in power plants, restricting non-power plant coal consumption, delineating prohibited areas for burning high-pollution fuels, and promoting the clean utilization of coal. Through the implementation of comprehensive coal reduction initiatives, it is anticipated that the proportion of coal consumption in primary energy consumption in Shijiazhuang City will decrease to below 45% by the year 2033.

Primary strategies for establishing a clean and low-carbon heating system encompass the expansion of centralized heat supply coverage, orderly promotion of centralized industrial heat supply, accelerated construction of clean energy heating sources, and the reinforcement of clean energy heating initiatives in rural areas. Notably, in 2020, the dispersion of coal combustion in rural plains of Shijiazhuang City was effectively eradicated. The envisioned goal for 2033 is the achievement of centralized heating and clean energy heating in all built-up areas at the county level and above in Shijiazhuang City.

Crucial measures to promote the use of renewable energy entail addressing challenges related to curtailed wind and solar power and actively fostering the construction of geothermal, wind, solar, and biomass energy utilization projects. Moreover, the encouragement of county-level biomass combined heat and power, biomass pellet boilers, and large-scale biomass natural gas projects in resource-appropriate areas is emphasized. By the year 2033, non-fossil energy consumption in Shijiazhuang City is projected to constitute over 18% of the total energy consumption.

Strategies for enhancing energy utilization efficiency involve the implementation of actions targeting both total energy consumption and intensity, robust energy conservation measures in key energy-consuming industries, and the active promotion of energy efficiency in construction. A targeted objective for 2033 is to increase the proportion of green urban construction areas in Shijiazhuang City to 80% of the newly constructed building area.

4.4.2 Realigning and Enhancing the Industrial Framework for a Green and Low-Carbon Industrial System

As per the pollution source emission inventory for Shijiazhuang City in 2017, an extensive network of more than 2500 industrial enterprises operates within a 30-kilometer radius centered on the main urban area of Shijiazhuang. In close proximity to the urban area, various industrial parks—including Zhengding New District, Circular Chemical Industrial Park, Liangcun Economic and Technological Development Zone, Airport Industrial Zone, and High-tech Industrial Development Zone—comprise an "industrial encirclement" layout, significantly influencing the air quality in the urban core of Shijiazhuang. Additionally, traditional industries are not highly concentrated in Shijiazhuang at present. Despite the articulation of dominant industries within industrial parks, some resident enterprises deviate from the functional positioning of these parks. Addressing this dynamic, there is a compelling need to expedite the relocation of polluting enterprises from urban zones to designated industrial parks, aiming to transform the industrial layout of Shijiazhuang. By the year 2033, the strategic objective entails achieving the relocation of key industrial enterprises in each county (city) within Shijiazhuang to specialized industrial bases or parks in consonance with their primary functions.

Within the enterprises above the designated size in Shijiazhuang City, the six major high-energy-consuming industries collectively constitute more than 85% of the industrial landscape. Traditional sectors, including the food industry, textile and apparel industry, petrochemical industry, steel industry, and building materials industry, command a substantial share within the industrial portfolio. While Shijiazhuang boasts a diverse spectrum of industrial categories, the prevailing industrial structure is characterized by a degree of extensiveness, manifesting as severe overcapacity in certain sectors. The development trajectory of traditional industries exhibits a disproportionate emphasis in comparison to the advancement observed in high-tech industries. Notably, when compared to neighboring provincial capitals or municipalities directly under the Central Government, Shijiazhuang registers the highest emission intensity per unit of GDP. Addressing this circumstance necessitates a dual-pronged approach. Firstly, stringent environmental access standards are rigorously implemented through the meticulous control of high-energy-consuming and high-polluting industries, vigilant oversight of relocation and upgrading projects, stringent limitations on the construction of new coal-fired projects, and heightened scrutiny of projects associated with VOCs. Simultaneously, there is a pronounced imperative to prioritize the industrial structure by reducing inefficient and excess capacity in industries such as steel, cement, coking, thermal power, casting (excluding precision casting), non-ferrous metals, carbon, calcium-magnesium, coal chemical, ceramics, and brick and tile.

4.4.3 Enhancing Pollution Mitigation in Key Sectors and Thoroughly Executing the Rectification of Volatile Organic Compounds

Shijiazhuang City boasts a considerable industrial scale, positioning its pollutant emissions among the forefront of the "2 + 26" cities. In the year 2017, the city ranked third, fourth, fifth, second, fourth, and fourth for total emissions of SO_2, NO_x, VOCs, NH_3, PM_{10}, and $PM_{2.5}$, respectively, within the "2 + 26" city classification.

The prevailing state of clean production in Shijiazhuang's industrial enterprises is presently suboptimal, necessitating the implementation of comprehensive environmental control measures. Central to these measures is the upgrading and transformation of non-power industries, including steel, coking, cement, glass, refractory materials, and ceramics. Additionally, specialized rectification efforts targeting industrial furnaces and kilns, as well as VOCs, are being implemented, alongside an overarching strategy for comprehensive governance within industrial zones. By the year 2033, the strategic vision involves key industrial enterprises in Shijiazhuang City achieving ultra-low emissions in both organized and unorganized operations, extending throughout the entirety of material transportation systems.

4.4.4 Proactively Restructuring the Transportation Framework and Advancing the Green and Low-Carbon Transportation System

The principal measures for restructuring the transportation system in Shijiazhuang encompass a substantial increase in the share of railway freight transportation, the robust development of multimodal transportation, expeditious adjustments to the logistics transportation structure, and the reinforcement of bypass road construction. Through targeted initiatives, such as incentivizing key enterprises in the steel and power sectors, large specialized logistics parks, and trading distribution centers, to either establish new or expand existing railway lines, and actively promoting the external opening and shared use of privately-built railway lines by enterprises, it is envisioned that the dedicated railway transportation proportion for key enterprises in the steel and power sectors will surpass 80% by the year 2033. Simultaneously, efforts are underway to accelerate airport and land port construction, foster the multimodal and seamlessly interconnected freight hubs (logistics parks) supported by railway logistics bases and road ports, and augment the collection and distribution channels surrounding Shijiazhuang Zhengding International Airport. Furthermore, there is a concentrated endeavor to comprehensively propagate the utilization of new energy logistics distribution vehicles and new energy operational vehicles within logistics parks, augment the construction of charging infrastructure within distribution centers, and establish a pilot project for environmentally friendly urban distribution.

The establishment of a green and low-carbon transportation system in Shijiazhuang City is primarily undertaken through the following three measures.

Foremost, there is an imperative to accelerate the development of green public transportation in Shijiazhuang. This encompasses the advancement of rail transit construction, expeditious planning and execution of bus rapid transit (BRT) systems, and the optimization of the layout of subway entrances, exits, and bus stops to achieve seamless connectivity within the public transportation network. By the year 2033, the anticipated contribution rate of motorized public transportation in Shijiazhuang is poised to surpass 70%. Additionally, exploring the planning of "zero-emission driving zones" in concentrated office areas or commercial zones is under consideration.

Subsequently, optimizing the composition of the existing vehicle fleets assumes paramount importance. By the specified timeline, the replacement of all public buses, sanitation vehicles, postal vehicles, taxis, commuter vehicles, and light-duty logistics delivery vehicles in Shijiazhuang with new energy vehicles is envisaged. Given the significant contribution of old vehicles and high-emission diesel vehicles to air pollutant emissions, there is a critical need to intensify efforts to phase out aging vehicles and enhance the treatment of heavy-duty diesel vehicles. By 2033, Shijiazhuang City is committed to the complete elimination of all diesel vehicles compliant with China IV vehicle emission standards and below, as well as gasoline vehicles adhering to China III vehicle emission standards and below. Diesel trucks meeting the stipulated criteria should undergo retrofitting with diesel particulate filters and selective catalytic reduction devices. Furthermore, a rigorous enhancement of environmental inspections for heavy-duty trucks is deemed imperative.

Finally, the promotion of pollution control over non-road mobile sources is crucial. Shijiazhuang City exhibits relatively high emissions from non-road mobile machinery. Primary measures for the prevention and control of air pollution from non-road mobile machinery involve designating and publicizing areas where the use of high-emission non-road mobile machinery is prohibited, implementing the fourth-stage emission standards for non-road mobile machinery, and actively promoting the transition from conventional fuel to electricity in non-road mobile machinery.

4.4.5 Optimizing and Restructuring Land Use Allocation while Advancing the Control of Non-point Source Pollution

In Shijiazhuang City, PM_{10} emissions originating from dust sources surpass those emanating from industrial and civil origins, necessitating heightened control measures targeting dust sources. Specifically, in the domain of construction dust, diverse construction sites, encompassing newly built and ongoing constructions, municipal infrastructure projects, demolition sites, roads, and water conservancy projects, are slated for comprehensive upgrades, based on adhering rigorously to the "Six 100%" requirements. An imperative focus is also placed on advancing the construction of prefabricated buildings, with the strategic vision of exclusively employing prefabricated methods for all newly constructed buildings in Shijiazhuang City that are conducive to such construction by the year 2033. Addressing road dust involves the implementation of effective control measures, including the promotion of the "cleaning by grams and deep cleaning" operational model, augmented maintenance of roads in urban-rural fringe areas, and the finalization of maintenance tasks for rural roads and the solidification of earth roads in each county (city, district). A crucial facet is the intensification of cleaning initiatives for roads in villages and towns. Shijiazhuang City is unwavering in its commitment to restricting the average dustfall to within 5 tons/(month·square kilometer) by 2025. Regarding industrial storage yards, a pivotal objective is the attainment of 100% coverage for online monitoring and video surveillance in key areas, including coal yards, material yards, and slag yards.

Shijiazhuang City, characterized by its extensive livestock and poultry farming, contends with substantial ammonia emissions from agricultural activities, necessitating the reinforcement of control measures against non-point source pollution in agriculture. Critical interventions include the prohibition of open burning of crop straws and the establishment of an enhanced system for the collection and storage of straws, with an emphasis on encouraging their resourceful utilization. Additionally, the imperative promotion of low-carbon agriculture is underscored, with Shijiazhuang City aiming to achieve near-zero emissions of agricultural waste by 2033, concurrently targeting a fertilizer utilization rate surpassing 50%.

Moreover, the restructuring of land use mandates the implementation of green carbon sink projects. This entails substantial expansion of urban green space areas, encompassing intensified efforts in mining site rehabilitation, plain greening initiatives, urban land hardening and re-greening, as well as the management of exposed land in rural areas. By the specified timeline, Shijiazhuang City aspires to attain a forest coverage rate exceeding 50%, accompanied by a greening and management rate surpassing 90% for exposed land within urban green spaces.

4.5 Analysis of Emission Reduction Potential

Drawing upon the atmospheric pollutant emission inventory of Shijiazhuang City in 2017 and the projected incremental emissions from motor vehicles and natural gas in 2033, coupled with the adjustment plans for energy, industrial, transportation, and land use structures, alongside end-of-pipe control measures for key industries, it is anticipated that by 2033, Shijiazhuang will achieve substantial reductions of 72.6%, 72.2%, 53.1%, 49.0%, and 71.7% in SO_2, NO_x, VOCs, NH_3, and primary $PM_{2.5}$ emissions, respectively, compared to 2017 levels.

A nuanced examination of various mitigation measures reveals the distinctive impact of each. Substitution with cleaner fuels emerges as the most influential contributor to SO_2 emissions reduction, constituting approximately 29.8% of the total emissions reduction. In the context of NO_x emissions reduction, the primary measure is the control of mobile source pollution, accounting for about 42.2% of the total emissions reduction. For VOCs emissions reduction, the measure with the highest contribution is the treatment of VOCs, representing approximately 33.3% of the total emissions reduction. NH_3 emissions reduction is predominantly influenced by agricultural non-point source pollution control, comprising around 89.7% of the total emissions reduction. Regarding $PM_{2.5}$ emissions reduction, the principal measure is resolving excess production capacity, making up roughly 26.0% of the total emissions reduction. Concerning CO_2 emissions reduction, the most significant contribution emanates from power structure adjustment and deep governance, constituting about 39.8% of the total emissions reduction (see Fig. 4.3).

4.6 Analysis of Air Quality Compliance

Utilizing the iterative calculation method for the atmospheric environmental capacity within the WRF-CMAQ air quality model, incorporating the simulated average proportions of sulfate, nitrate, and primary $PM_{2.5}$ to $PM_{2.5}$ in Shijiazhuang, emission

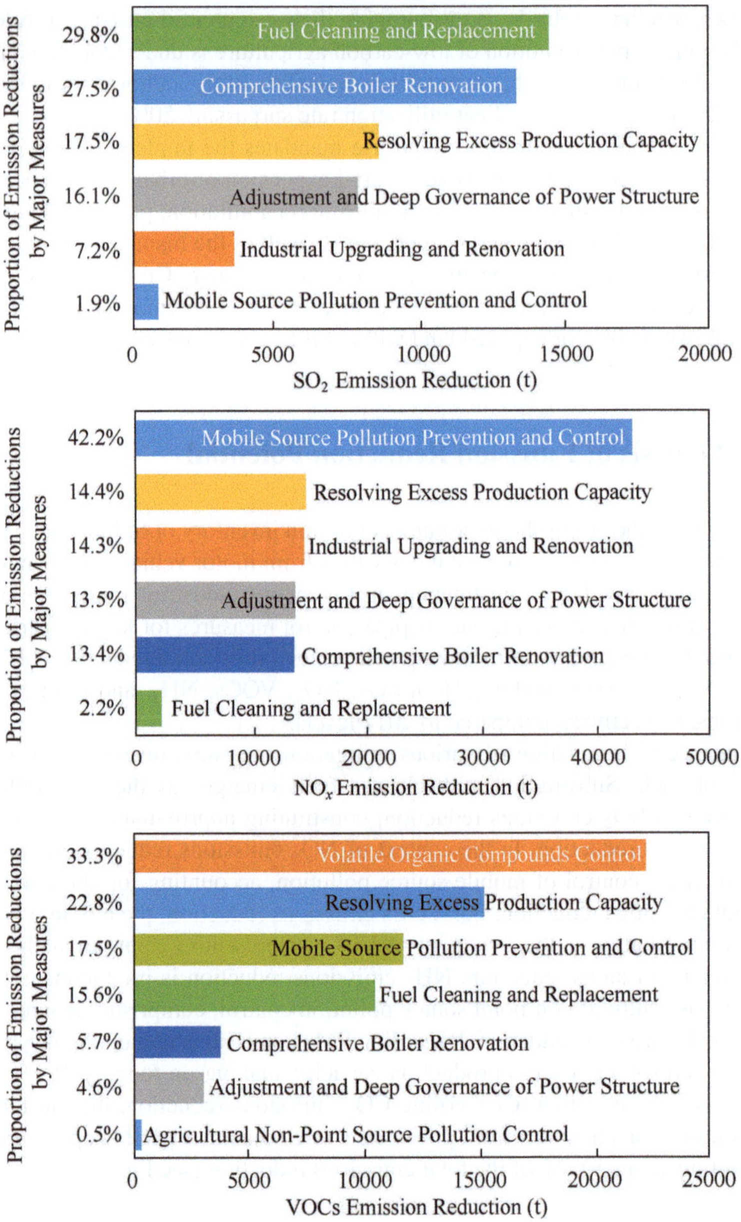

Fig. 4.3 Evaluation of the impact of key measures on reduction of atmospheric pollutants and CO_2 emissions

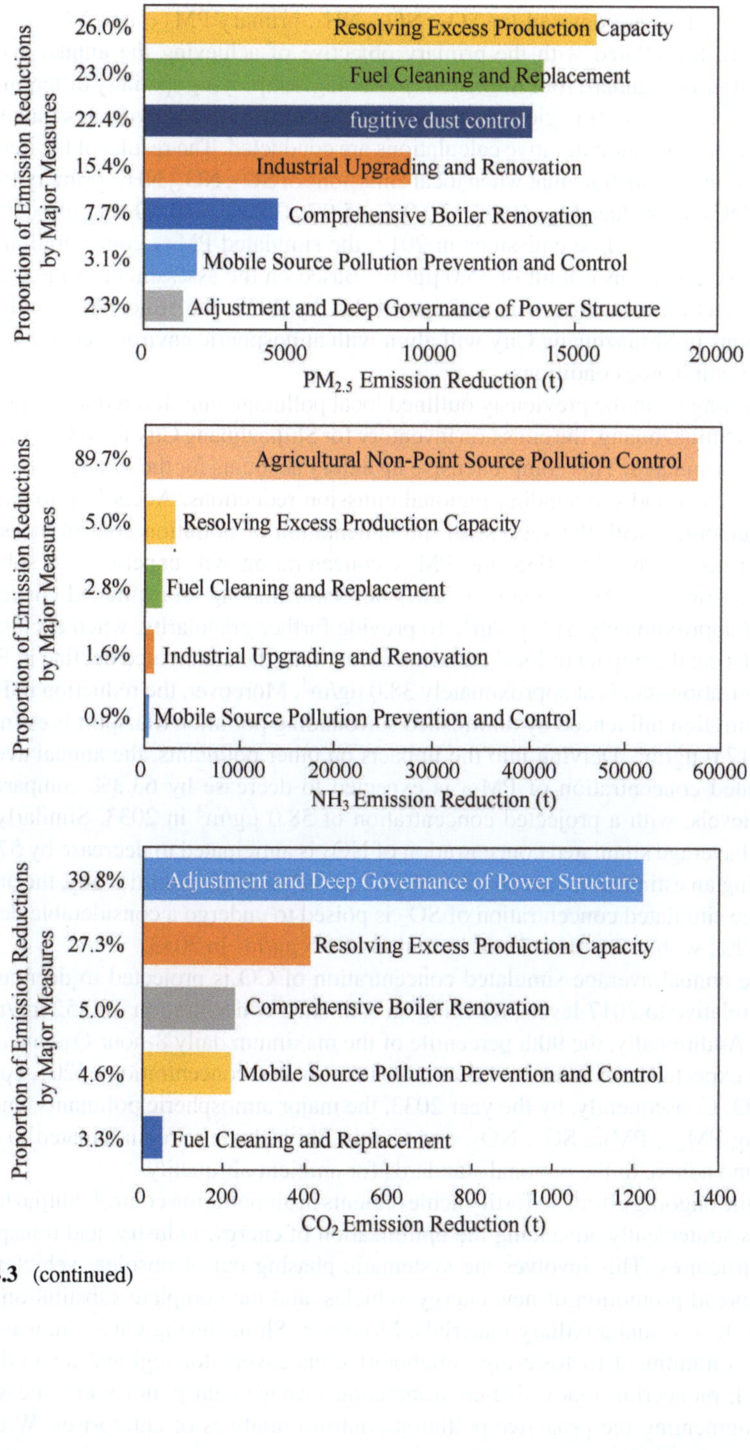

Fig. 4.3 (continued)

reduction plans are devised for SO_2, NO_x, NH_3, primary $PM_{2.5}$, and VOCs. These plans are formulated with the primary objective of achieving the annual average concentration standard for $PM_{2.5}$ in Shijiazhuang. Employing spatially differentiated emission reduction strategies, new emission inventories are iteratively generated, and numerical simulation iterative calculations are conducted. The results of the iterative simulation demonstrate that when local emissions of SO_2, NO_x, NH_3, primary $PM_{2.5}$, and VOCs are reduced by 60.0%, 70.0%, 45.0%, 70.0%, and 50.0%, respectively, relative to the baseline emissions in 2017, the simulated $PM_{2.5}$ concentration falls below the compliance limit of 35.0 μg/m^3. Based on the assessment of these emission reduction measures, it is anticipated that by 2033, the atmospheric pollutant emissions in Shijiazhuang City will align with atmospheric environmental capacity under compliance conditions.

Building upon the previously outlined local pollutant emission reduction proportions in Shijiazhuang, the emission inventory for Shijiazhuang City in 2033 is meticulously formulated. This comprehensive inventory accounts for the synergistic effects of both local and surrounding regional emission reductions. According to simulation outcomes, with the successful implementation of pollution control measures, it is projected that by 2033, the $PM_{2.5}$ concentration will experience a substantial reduction of 63.9% relative to 2017 levels, achieving an estimated concentration of approximately 31.0 μg/m^3. To provide further granularity, when exclusively considering the impact of local emission reductions, the anticipated decline in $PM_{2.5}$ concentration stands at approximately 38.0 μg/m^3. Moreover, the reduction in $PM_{2.5}$ concentration influenced by diminished surrounding pollution transport is estimated to be 17.0 μg/m^3. Delving into the impacts on other pollutants, the annual average simulated concentration of PM_{10} is expected to decrease by 63.3% compared to 2017 levels, with a projected concentration of 58.0 μg/m^3 in 2033. Similarly, the annual average simulated concentration of NO_2 is anticipated to decrease by 57.1%, reaching an estimated concentration of 23.2 μg/m^3 in 2033. Additionally, the annual average simulated concentration of SO_2 is poised to undergo a considerable decline of 80.0%, with an estimated concentration of 6.5 μg/m^3 in 2033.

The annual average simulated concentration of CO is projected to decrease by 4.1% relative to 2017 levels, reaching an estimated concentration of 0.52 mg/m^3 in 2033. Additionally, the 90th percentile of the maximum daily 8-hour O_3 concentration is expected to increase by 1.5%, with an estimated concentration of 201.0 μg/m^3 in 2033. Consequently, by the year 2033, the major atmospheric pollutants, encompassing $PM_{2.5}$, PM_{10}, SO_2, NO_2, and CO, in Shijiazhuang are anticipated to align with and adhere to the national standards for ambient air quality.

In the ongoing efforts to fortify achievements in air pollution control, Shijiazhuang City is strategically advancing the optimization of energy, industry, and transportation structures. This involves the systematic phasing out of obsolete vehicles, the widespread promotion of new energy vehicles, and the complete substitution with low-VOC raw and auxiliary materials. Moreover, Shijiazhuang City is unwavering in its commitment to fostering collaborative endeavors for regional air pollution control, pioneering innovative environmental management policies and measures, and augmenting the proactive pollution control initiatives of enterprises. With the implementation of these multifaceted initiatives, it is anticipated that the O_3 concentration in Shijiazhuang will conform to the Grade II national standards for air quality by the year 2035.

4.7 Analysis of Predictions for Peak Carbon Emissions

The long-term carbon emission scenarios, as simulated and predicted by the LEAP model for Shijiazhuang City, are visually depicted in Fig. 4.4. This comprehensive illustration incorporates the indirect carbon emissions associated with externally purchased electricity, based on the emission inventory. In the BAU scenario, there is marginal improvement in the energy structure and energy intensity across various sectors. Nevertheless, with the overarching adjustment in the industrial structure and a reduction in the proportion of the secondary industry, the growth rate of carbon emissions is anticipated to decelerate, ultimately reaching its peak around 2030. Contrastingly, in the AQP scenario, with the explicit goal of achieving air quality compliance by 2033, carbon emissions are projected to peak in 2020 at 95.6 million tons, followed by a gradual decline. In comparison to the BAU scenario, the carbon emissions in Shijiazhuang City are expected to decrease by 8.46 million tons, 13.48 million tons, and 15.82 million tons in the years 2025, 2030, and 2035, respectively, under the AQP scenario.

In the AQP scenario, the sector-wise breakdown of carbon emissions in Shijiazhuang City is detailed in Fig. 4.5.

By the year 2033, the industrial sector is poised to achieve a substantial reduction of 13.406 million tons in carbon emissions compared to the baseline scenario. In the industrial end-use sector, pivotal contributors to carbon emissions encompass steel, chemicals, and cement. Particularly noteworthy is the pivotal role of the steel sector in determining the total carbon emissions trend, given its substantial reliance on coal and coke. With the implementation of future capacity reduction policies, the steel production undergoes a gradual decline, resulting in a progressive reduction in

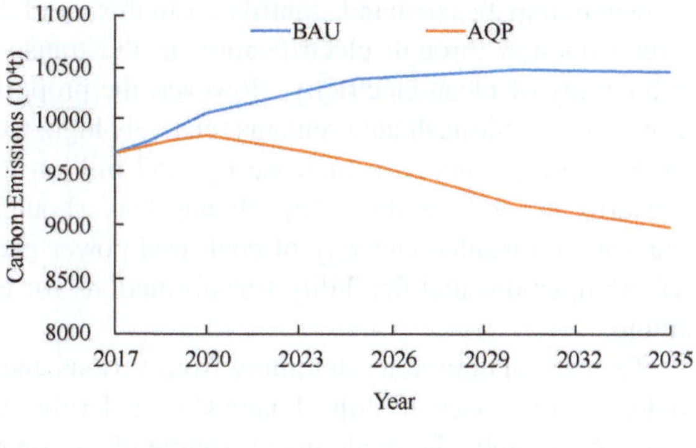

Fig. 4.4 Long-term carbon emissions scenarios for Shijiazhuang City simulated and predicted using the LEAP model

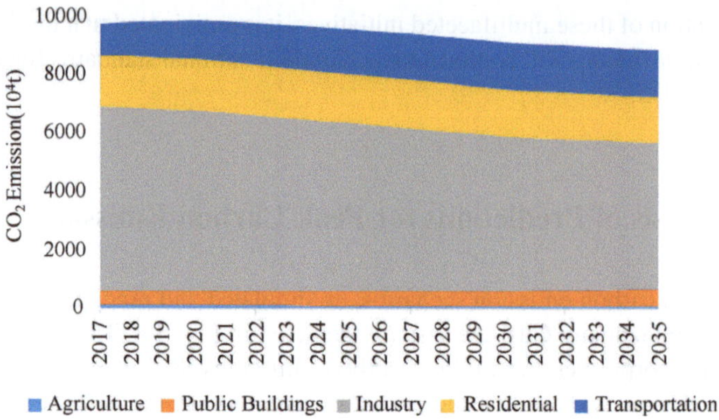

Fig. 4.5 Carbon emissions by sector in Shijiazhuang City under the AQP scenario

carbon emissions from the industrial end-use sector since 2017. Turning to the transformation sector, the AQP scenario outlines a phased reduction in externally supplied electricity, addressing air quality concerns associated with the "coal-electricity encirclement". Moreover, there is an anticipated increase in the share of solar and other clean energy sources, exceeding 15% by 2035. The total carbon emissions from the power sector are projected to peak around 2030. In aggregate, carbon emissions from the industrial sector in Shijiazhuang City peaked in 2018 and have exhibited a continuous decline since then.

The energy demand and carbon emissions from the commercial and civil building sectors in Shijiazhuang have not reached their zenith. However, proactive measures such as the proscription of bulk coal use in residential areas and the promotion of centralized heating are expected to result in a significant enhancement of energy efficiency. This initiative is foreseen to contribute to a noteworthy reduction of 2.79 million tons in carbon emissions by the year 2033, relative to the baseline scenario.

The overall energy demand and carbon emissions in the transportation sector of Shijiazhuang have not yet reached their peak, driven by two primary factors. Firstly, the continuous growth in the number of private cars, coupled with sustained growth in logistics demand, contributes to this trend. Secondly, the success of emission reduction through electrification in the transportation sector hinges on the availability of clean electricity. However, the proportion of local coal-fired power generation in Shijiazhuang remains relatively high. Consequently, carbon emissions in 2033 are projected to increase by 1.57 million tons compared to the baseline scenario. To address this, Shijiazhuang City should adopt stringent measures to control the installed capacity of coal-fired power generation and expedite energy-saving upgrades and flexibility transformations for existing coal-fired units in the future.

The carbon emissions stemming from various energy consumption patterns and industrial processes in Shijiazhuang City under the AQP scenario are delineated in Fig. 4.6. Notably, the predominant source of carbon emissions related to industrial

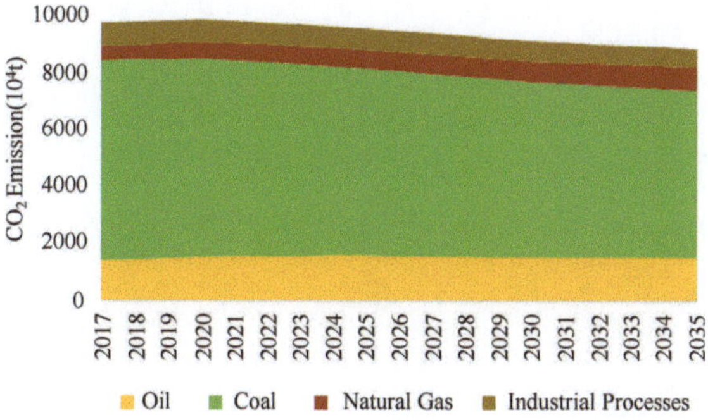

Fig. 4.6 Carbon emissions from diverse energy sources and industrial processes in Shijiazhuang City under the AQP scenario

processes is the cement production process. On the whole, there is a slight downward trajectory in carbon emissions originating from industrial processes. Carbon emissions resulting from coal consumption manifest a substantial decrease, whereas those arising from natural gas usage demonstrate an incremental year-on-year increase. This observed trend is primarily ascribed to the diminishing reliance on coal and the shift in end-use energy preferences in both industrial and civil domains, transitioning from coal to natural gas. Anticipated carbon emissions from petroleum consumption are poised to reach their peak around 2025, followed by a gradual reduction. This decline is attributed to advancements in the transportation structure within the freight sector and the widespread adoption of electrification in motor vehicles.

Chapter 5
Analysis of the Pathways to Achieving "Dual Attainments" in Huzhou City

Abstract Within this chapter, Huzhou City is employed as a case study to scrutinize the scenario of "dual attainments" within the realm of high-tech cities. Through an in-depth examination encompassing the general landscape of Huzhou City, the prevailing environmental air quality conditions, the distinctive features of atmospheric pollutant emissions, and the established deadlines for achieving air quality compliance, pathways are delineated for Huzhou City to meet air quality standards. These pathways are informed by comprehensive analyses derived from urban emission inventories and model simulations. Additionally, the chapter concurrently explores the scenario of peak carbon emissions within the context of achieving air quality standards.

Model calculations suggest that achieving air quality compliance in Huzhou City necessitates notable reductions in pollutant emissions compared to 2017 baseline levels. Specifically, reductions of 26.7%, 28.6%, 10%, 28.9%, and 20% respectively for SO_2, NO_x, NH_3, primary $PM_{2.5}$, and VOCs are required, which results in simulated $PM_{2.5}$ concentrations falling below 35.0 μg/m^3. Furthermore, simulations indicate that by 2020, the annual average concentrations of $PM_{2.5}$, PM_{10}, SO_2, CO, and NO_2 will align with regulatory standards. These concentrations are estimated to reach 32.3 μg/m^3, 45.6 μg/m^3, 13 μg/m^3, 1.2 mg/m^3, and 31.4 μg/m^3, respectively. Despite a 3.1% decrease compared to the 2017 baseline level, the 90th percentile of the maximum daily 8-hour O_3 concentration is anticipated to exceed the standard. However, it is projected that through further coordinated measures aimed at reducing VOCs emissions, O_3 levels will attain compliance by 2025. Regarding peak carbon emissions, in the scenario where air quality in Huzhou City achieves compliance by 2020, initiatives such as the closure of several cement and brick factories during 2017–2020 have resulted in a peak in carbon emissions in 2017. Subsequently, a stabilization of emissions is expected, followed by a gradual decline post-2028.

Y. Bo et al., *Pathways for Reaching Air Quality Standard and Carbon Peak in Selected Cities in China*, Urban Sustainability, https://doi.org/10.1007/978-981-96-8315-4_5

5.1 Overview of Huzhou City

Huzhou City is situated in the northern region of Zhejiang Province, contiguous to Jiaxing City to the east and adjoining Hangzhou City to the south. In terms of administrative division, the city encompasses two districts and three counties, spanning a total area of 5,820 square kilometers. In 2017, the permanent population of Huzhou City reached 2.995 million, reflecting an urbanization rate of 62.0%. The distribution of economic activities across primary, secondary, and tertiary sectors stood at 5.1%, 47.4%, and 47.5%, respectively. Notably, coal consumption constituted 49.6% of the city's overall energy consumption. Furthermore, the aggregate motor vehicle ownership surpassed 734,000 units.

Table 5.1 presents the annual pollution indicators for various pollutants in Huzhou City spanning the years 2013 to 2017. Notably, the annual average indicators for PM_{10}, NO_2, SO_2, and CO in Huzhou City have already achieved compliance with Grade II national standards for air quality. However, despite these accomplishments, Huzhou City encountered discernible challenges in meeting air quality standards by the conclusion of 2017. In the year 2017, the annual average concentration of PM2.5 in Huzhou City reached 42.2 $\mu g/m^3$, surpassing the prescribed limit of the Grade II national standards for air quality by approximately 20.6%. Moreover, the 95th percentile of the daily PM2.5 concentration escalated to 95.0 $\mu g/m^3$, exceeding the 24-hour average concentration limit of the Grade II national standards for air quality by about 26.7%. Concurrently, the 90th percentile of the maximum daily 8-hour O_3 concentration recorded 186.6 $\mu g/m^3$, surpassing the stipulated limit by 16.6%. Despite a decreasing trend observed in the annual average concentration of $PM_{2.5}$ over the preceding 5 years, it continues to fall short of compliance, necessitating heightened efforts in pollution control. Furthermore, the persistently higher-than-standard levels of O_3 concentration in Huzhou City pose a substantial challenge, significantly impacting the proportion of days when air quality aligns with prescribed criteria. This underscores the imperative for targeted control measures.

Table 5.1 Annual pollution indicators for various pollutants in Huzhou City from 2013 to 2017

Pollutants	2013	2014	2015	2016	2017
PM2.5	74.2	64.2	57.6	46.9	42.2
PM_{10}	110.6	86.9	75.6	68.9	63.9
O_3_8h_90per	181.0	166.0	189.5	195.0	186.6
NO_2	52.0	48.2	40.0	36.5	38.2
SO_2	28.5	21.8	16.8	17.4	14.7
CO_95per	1.9	1.4	1.6	1.3	1.3

Notes:
1. O_3_8h_90per signifies the 90th percentile of the maximum daily 8-hour O_3 concentration, while CO_95per indicates the 95th percentile of daily average CO concentration
2. $PM_{2.5}$, PM_{10}, O_3_8h_90th, NO_2, and SO_2 concentrations are stated in $\mu g/m^3$, whereas CO_95per concentration is articulated in mg/m^3

Considering the prevailing air quality conditions in Huzhou City and the available measures for corrective action, the city has instituted air quality compliance objectives in alignment with pertinent strategic frameworks such as the "13th Five-Year Ecological and Environmental Protection Plan of Zhejiang Province" and the "13th Five-Year Energy Development Plan of Huzhou City". In elucidation, by the year 2020, the aspiration is for the $PM_{2.5}$ concentration to substantially conform to the Grade II national standards for ambient air quality. Simultaneously, the concentrations of PM_{10}, SO_2, CO, and NO_2 are targeted to consistently meet the requisites of the Grade II national standards for ambient air quality. Looking further ahead, by 2025, it is envisaged that the O_3 concentration will align with the Grade II national standards for ambient air quality.

5.2 Analysis of the Emission Characteristics for the Baseline Year in Huzhou City

Figure 5.1 provides a detailed breakdown of the contribution rates of atmospheric pollutants and CO_2 emissions in Huzhou City for the year 2017. Notably, process sources emerge as the predominant contributors to emissions of SO_2, NO_x, PM_{10}, $PM_{2.5}$, and VOCs in Huzhou City, constituting 55.0%, 42.2%, 48.6%, 32.0%, and 35.3% of the total emissions, respectively. NH_3 emissions in the city predominantly originate from agricultural sources, accounting for 63.7% of the total emissions. In the realm of CO_2 emissions, stationary combustion sources take precedence, contributing 77.0% to the overall emissions.

5.3 Forecast of Economic and Energy Development

5.3.1 Forecast of Economic Development

According to the "Overall Urban Planning of Huzhou (2017–2035)" and recent population growth trends, the permanent population in Huzhou is forecasted to reach 3.037 million in 2020. Factoring in the influence of economic growth, shifts in industrial structure, and economic development efficiency on population dynamics, the estimated urbanization rate in Huzhou is approximately 63.8%, which is determined through linear interpolation. Analysis derived from this urbanization rate projection suggests that the urban population in Huzhou City is expected to be around 1.938 million in 2020. Utilizing historical data extrapolation, the annual average growth rate of Huzhou City's GDP from 2018 to 2020 is calculated at 6.8%. This figure, combined with the annual average growth rate of Huzhou City's GDP from 2008

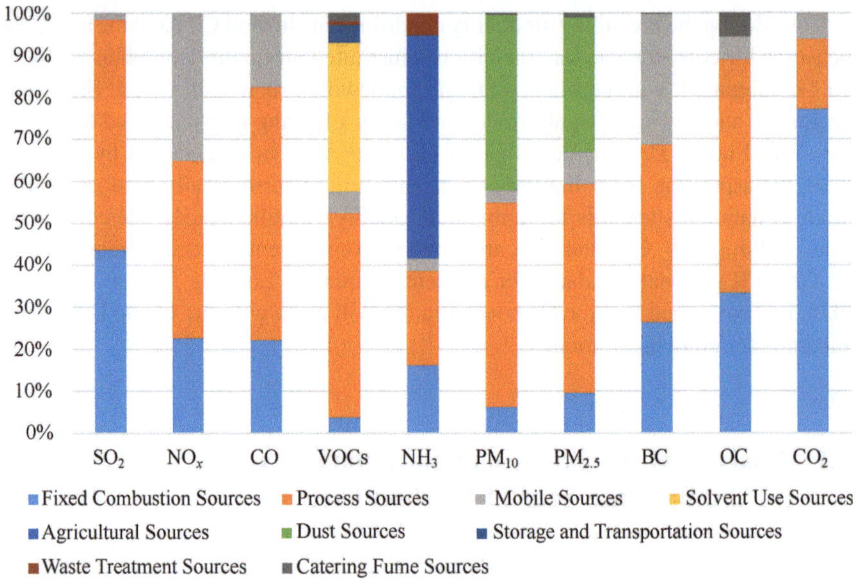

Fig. 5.1 Contribution rates of atmospheric pollutants and CO_2 emissions in Huzhou City in 2017

to 2017, as documented in the "Huzhou Statistical Yearbook," leads to an anticipated GDP of 284.28 billion yuan for Huzhou City in 2020 by utilizing the elasticity coefficient method.

5.3.2 *Forecast of Energy Development*

In accordance with the "13th Five-Year Energy Development Plan of Huzhou City", there will be a concerted effort in Huzhou to promote the utilization of clean energy, particularly natural gas. By the year 2020, the total energy demand in Huzhou City is anticipated to reach approximately 12.99 million tons of standard coal. Notably, coal consumption is forecasted to decrease to around 8 million tons, marking a decline from 49.6% of primary energy consumption in 2017 to approximately 40.0% in 2020. Concurrently, the proportion of natural gas consumption in primary energy consumption is projected to rise from 6.9% in 2017 to approximately 10% in 2020. This entails a total natural gas consumption of 1,070 million cubic meters in 2020, representing an increase of 438.53 million cubic meters from the 2017 levels. Furthermore, the proportion of non-fossil energy consumption in primary energy consumption is expected to increase from 16.9% in 2017 to about 18% in 2020, indicative of a substantial rise in the share of clean energy sources (refer to Fig. 5.2).

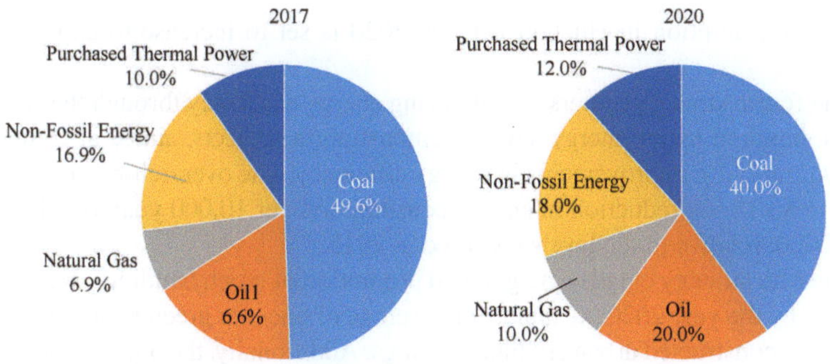

Fig. 5.2 Energy structure in Huzhou City for the years 2017 and 2020

5.4 Key Tasks and Measures for Emission Reduction

5.4.1 Expediting the Transition of the Energy Structure and Establishing a Clean and Low-carbon Energy System

In 2017, coal consumption accounted for 49.6% of the primary energy consumption in Huzhou City, and enterprises above the designated size were responsible for 83.6% of the overall coal consumption. Given the limited share of clean energy sources, there is an urgent need to accelerate the restructuring of the energy sector.

Outlined below are the strategies aimed at adjusting the energy structure.

The primary strategy involves stringent control over total coal consumption, with the objective of decreasing the proportion of coal consumption in primary energy consumption to below 40% by the year 2020 in Huzhou City.

The second strategy involves the proactive elimination of fuel facilities with high pollution. By the conclusion of 2019, Huzhou City aims to substantially phase out coal-fired boilers with a steam capacity of 35 steamed tons per hour or below and to proscribe coal-fired hot blast stoves. In the non-ferrous metal industry, the successful removal of coal-fired drying kilns, coal-fired reverberatory furnaces, and coal-fueled lead-melting pots is anticipated. Furthermore, coal-fired heating and drying furnaces (kilns) within the cogeneration heating network's coverage area are expected to undergo substantial eradication, and there will be a vigorous elimination of fuel gas generators with furnace diameters below 3 m.

The third strategy entails the augmentation of clean energy utilization. There is a critical need to expedite the expansion of natural gas supply capacity, with the objective of achieving a representation of 10% for natural gas consumption in the city's primary energy consumption by the year 2020. Furthermore, it is imperative to actively foster the large-scale utilization of non-fossil energy sources, such as photovoltaics and geothermal energy. The target for the proportion of non-fossil energy consumption in Huzhou City by 2020 is set to increase to approximately 18%.

The fourth strategy centers on enhancing energy efficiency through the execution of one hundred major energy-saving demonstration projects, aimed at comprehensively promoting improvements in energy efficiency. The overarching objective is to achieve an 18.5% reduction in energy consumption per 10,000 yuan of GDP by the year 2020, relative to the levels recorded in 2015.

The fifth strategy entails a heightened promotion of green buildings. Huzhou City aspires, by the year 2020, to attain full coverage of one-star green building standards for newly constructed urban civilian buildings. Additionally, the objective is to have a proportion exceeding 20% for buildings meeting a two-star or higher green building standard.

5.4.2 *Adjusting and Optimizing the Industrial Structure for a Green and Low-carbon Industrial System*

In 2017, Huzhou City exhibited an industry structure characterized by proportions of 5.1: 47.4: 47.5 for the primary, secondary, and tertiary sectors, respectively. This distribution suggests a relatively balanced proportion between the secondary and tertiary sectors, with the primary sector holding a smaller share. Notably, industrial energy consumption in Huzhou City was concentrated within eight high-energy-consuming industries, contributing 42.6% to the total added value from industries above the designated size and accounting for 75.8% of the total energy consumption in industries above the designated size. Given these circumstances, there is a pressing need for Huzhou City to expedite the environmentally conscious transformation of the existing industrial system. This involves the vigorous development of emerging industries and high-end manufacturing characterized by low energy consumption and high added value. The strategic focus should also prioritize the promotion of the low-carbon transformation within traditional manufacturing industries. The envisioned transformation and upgrading of the building materials and wood industries by the end of 2020 will be comprehensively achieved through the implementation of measures such as closures and elimination, park management, and standardized improvements. Concurrently, the standards for process equipment, product quality, pollution control, and carbon emissions are anticipated to align with domestic advanced standards.

On one hand, stringent environmental access controls are imperative. Industries including steel, coking, electrolytic aluminum, casting, cement, and flat glass are explicitly prohibited from augmenting new production capacities. In instances of relocation or reconstruction projects, a mandatory two-fold reduction in pollutant emissions should be enforced as an alternative measure. Concurrently, the environmental access threshold for industries identified as key VOC emitters should be elevated to regulate the emission of new VOCs.

On the other hand, the transformative efforts target high-pollution industries, including building materials, coatings, casting, enameled wire, chemicals, chemical fiber, printing, dyeing, pianos, woodworking, and plastic products. The emphasis is on expediting the elimination of outdated capacity, processes, and equipment characterized by excessive energy consumption and severe pollution. By the conclusion of 2018, Huzhou City aims to cease operations of 16 clay brick and tile enterprises, along with one cement grinding production enterprise. Furthermore, it is anticipated that by the end of 2019, one cement grinding production enterprise and three cement clinker production lines will be closed in Huzhou City.

5.4.3 Intensifying Pollution Control in Key Industries and Systematically Executing the Remediation of VOCs

The overall adherence to clean production standards within industrial enterprises in Huzhou City is notably suboptimal. Addressing this concern requires an urgent establishment of a comprehensive "one factory, one strategy, one file" system. This system should leverage online monitoring data for flue gas as a foundation for law enforcement, wherein penalties for exceeding standards are heightened, and collaborative disciplinary efforts are intensified. The overarching objective is to attain full compliance with emissions from industrial pollution sources. To comprehensively address pollution in key industries, several primary measures are implemented: (1) Executing thorough control over coal-fired power plants. (2) Strengthening boiler smoke emission standards. (3) Elevating the level of exhaust gas treatment in key industries. (4) Undertaking special actions to rectify industrial furnaces. (5) Implementing specific treatments for VOCs. (6) Advancing the comprehensive treatment of waste gases in key industrial parks. (7) Reinforcing control over unorganized emissions from industrial enterprises.

5.4.4 Proactively Adjusting the Transportation Structure and Advancing the Green and Low-carbon Transportation System

In 2018, the anticipated road freight volume and waterway freight volume in Huzhou are expected to constitute 61.2% and 38.8% of the total freight volume, respectively, underscoring the continued dominance of road transportation in the movement of goods. The principal strategy for the adjustment of the transportation structure in Huzhou involves leveraging the pivotal roles of waterways and railways in the long-distance transportation of bulk commodities. This strategy aims to augment the proportion of waterway and railway freight, foster the development of multimodal transport, and establish trunk-branch connection freight hubs (logistics parks) supported by inland river ports, road ports, and railway logistics bases. A simultaneous focus is directed towards expediting the promotion of container multimodal transport. By the conclusion of 2018, Huzhou City aspires to institute five multimodal transport and trunk-branch connection freight hubs (logistics parks). Additional endeavors will concentrate on promoting the containerization of suitable containerized goods, thereby increasing the proportion of transportation in alignment with the paradigm of "shift from land to water and from bulk to container."

The construction plan for the green and low-carbon transportation system in Huzhou City outlines the following strategies. The "public transportation priority" initiative will be actively pursued, with a targeted increase in the contribution rate of motorized public transportation to over 20% by the year 2020. Additionally, there will be a proactive promotion of new energy vehicles, aiming for a 100% proportion of new energy or clean energy vehicles in newly added or updated public buses, sanitation vehicles, postal vehicles, taxis, commuter vehicles, and light logistics delivery vehicles. Specifically, by the conclusion of 2018, all public buses in the urban area of Huzhou City are slated to achieve 100% electrification. Furthermore, by the end of 2020, the utilization rate of new energy or clean energy vehicles in the taxi fleet of Huzhou is projected to reach 94%.

Simultaneously, there is an imperative to enhance the comprehensive management of in-use motor vehicles and vessels. Firstly, the elevation of emission standards for motor vehicles is deemed essential, necessitating the implementation of the sixth phase of emission standards ahead of schedule. This entails promoting the use of gas vehicles compliant with China VI vehicle emission standards. Secondly, the expeditious phase-out of old vehicles is crucial, with the targeted objective of essentially eliminating diesel vehicles adhering to China III vehicle emission standards and below by 2025. Thirdly, the governance of heavy-duty trucks demands intensified efforts, encompassing the establishment of an integrated and comprehensive monitoring system labeled "sky, earth, vehicle, and people." This involves the implementation of emission testing and mandatory maintenance systems for in-use vehicles, along with the promotion of retrofitting diesel particulate filters and selective catalytic reduction devices. Fourthly, in addressing non-road mobile machinery, it is imperative to research and establish a dynamic updating system for emission inventories of non-road mobile machinery. This includes designating low-emission control areas. Additionally, expediting the application and promotion of clean energy sources such as hybrid, pure electric, and fuel cell technologies in non-road mobile machinery is deemed necessary. Fifthly, concerning vessels, there is a pressing need to actively guide water transport operators in retiring inland river shipping vessels with a service life exceeding 20 years. Simultaneously, there is a proactive push to promote the adoption of new energy or clean energy sources such as electricity and natural gas for vessels. The overarching goal is to achieve nearly complete coverage of power in inland river port shores by the year 2020.

5.4.5 *Optimizing and Adjusting Land Use Structure and Advancing the Control of Non-point Source Pollution*

The principal endeavors for optimizing and adjusting land use structure in Huzhou City involve reinforcing dust management, executing measures for agricultural non-point source pollution control, and augmenting green carbon sinks.

5.4.5.1 Strengthening Dust Management

The enhancement of construction dust control is chiefly centered on proactive measures such as the establishment of green construction sites, the comprehensive implementation of the "Seven 100%" long-term mechanism, and the vigorous promotion of prefabricated building construction. A targeted objective is to ensure that the proportion of prefabricated buildings in newly constructed buildings throughout Huzhou City is no less than 30% by the end of 2020.

In terms of road dust management, the primary focus is on strictly adhering to cleaning quality standards and expediting the mechanized cleaning of roads. The goal is to achieve a 100% mechanized cleaning rate for expressways and national/provincial highways, while maintaining the average dustfall in each county (district) below 5 tons/(month·square kilometer) by 2020.

The management of dust emissions in storage yards predominantly involves the comprehensive initiation of windproof and dust suppression facility construction in all coal storage yards and coal unloading sites. Specifically, the ground hardening rate of inland river terminals prone to dust is expected to reach 100%, and the coverage rate of dust suppression facilities in storage yards is anticipated to achieve 100%.

5.4.5.2 Implementing Agricultural Non-point Source Pollution Control

Primarily, it is imperative to proscribe the open burning of straw, institute a grid-based supervision mechanism for the prohibition of straw burning, and comprehensively advance the industrialization of straw utilization. By 2020, the envisaged outcome is an overall straw utilization rate in Huzhou City surpassing 95%.

Secondly, the focus should be on enhancing the resource utilization of livestock and poultry manure. The completion of the rectification of pig farms with an inventory of over 500 heads is targeted by the end of 2018. Simultaneously, there is a critical need to vigorously promote the reduction and efficiency enhancement of chemical fertilizers and pesticides in agriculture. By 2020, the aspiration is for a 6% decrease in the usage of chemical pesticides compared to 2017 levels, and the fertilizer utilization rate is aimed to exceed 40%.

5.4.5.3 Enhancing Green Carbon Sinks

Primarily, prioritizing the expeditious establishment of the National Green Mining Development Demonstration Zone holds paramount significance. The target is to achieve a 100% completion rate of green mines throughout the entire city by 2020. Additionally, there is a crucial emphasis on advancing the “mine land reclamation” special action, with all associated tasks slated for completion by the end of 2019.

Secondly, the imperative is to sustain afforestation efforts. This involves intensifying the establishment of forest cities and towns, creating urban greenways and green corridors, and executing the initiative for “returning industrial land to forests”. These measures significantly contribute to enhancing green coverage in urban built-up areas. By 2020, Huzhou City aspires to attain a forest reserve of 4.2 million acres, with a greening rate exceeding 29% for plain areas and a forest coverage rate of 48.5%.

5.5 Analysis of Emission Reduction Potentials

In accordance with the atmospheric pollutant emission inventory of Huzhou City in 2017 and the anticipated incremental emissions from motor vehicles and natural gas in 2020, coupled with adjustment plans for energy, industrial, transportation, and land use structures, and end-of-pipe control for key industries in Huzhou City, it is prognosticated that by 2020, there will be noteworthy reductions of 28.0%, 33.8%, 38.5%, 12.0%, and 39.9% in SO_2, NO_x, VOCs, NH_3, and primary $PM_{2.5}$ emissions in comparison to 2017 levels, respectively.

Upon comparing various measures, it is evident that the elimination of cement production capacity and the retrofit for ultra-low emissions emerge as the most impactful contributors to SO_2 emissions reduction, constituting approximately 40.4% of the total emissions reduction. For NO_x emissions reduction, the predominant measure is the elimination of cement production capacity and the retrofit for ultra-low emissions, accounting for about 65.2% of the total emissions reduction. In the domain of VOC emissions reduction, the measure with the highest contribution is the treatment of VOCs, representing approximately 71.5% of the total emissions reduction. Regarding $PM_{2.5}$ emissions reduction, the primary measure is comprehensive dust control, making up roughly 47.9% of the total emissions reduction. NH_3 emissions reduction is predominantly influenced by agricultural non-point source pollution control, comprising around 85.7% of the total emissions reduction. Concerning CO_2 emissions reduction, the most significant contribution emanates from the elimination of cement production capacity, making up about 65.1% of the total emissions reduction (refer to Fig. 5.3).

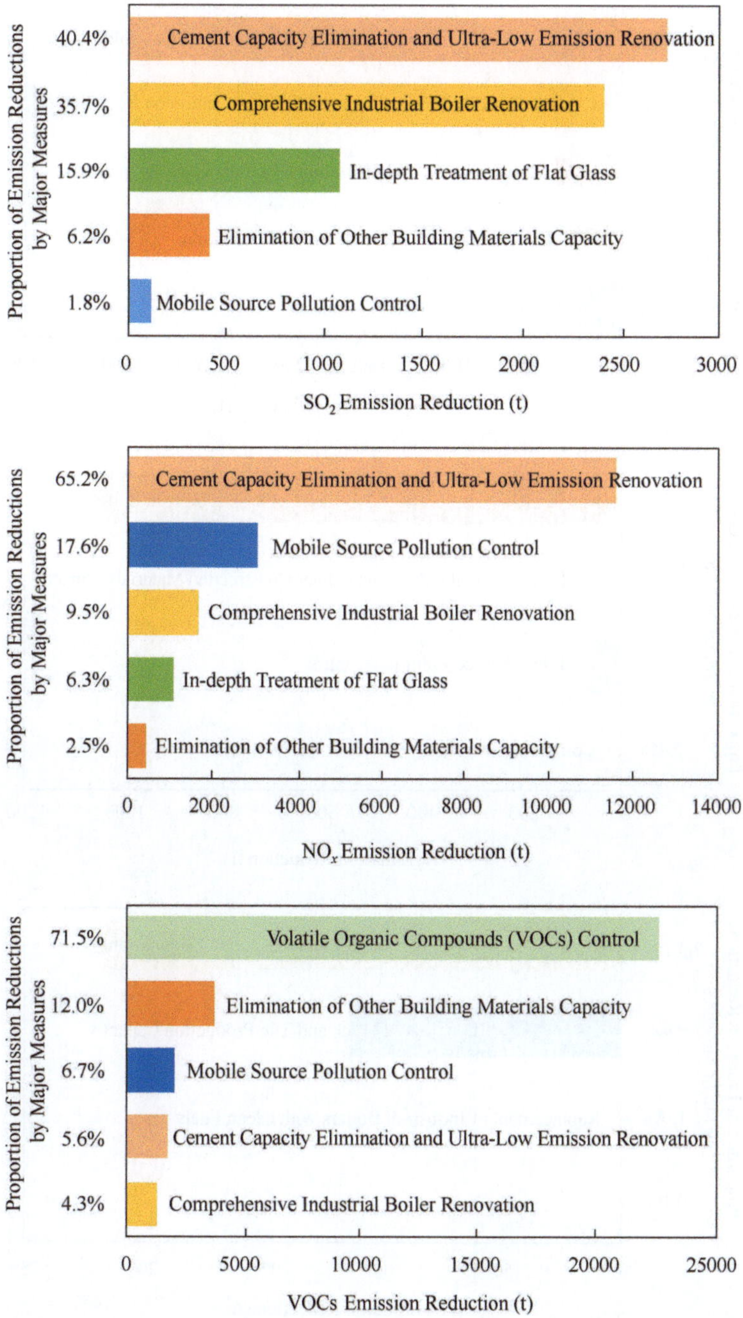

Fig. 5.3 Evaluation of the impact of principal measures on the reduction of atmospheric pollutants and CO_2 emissions

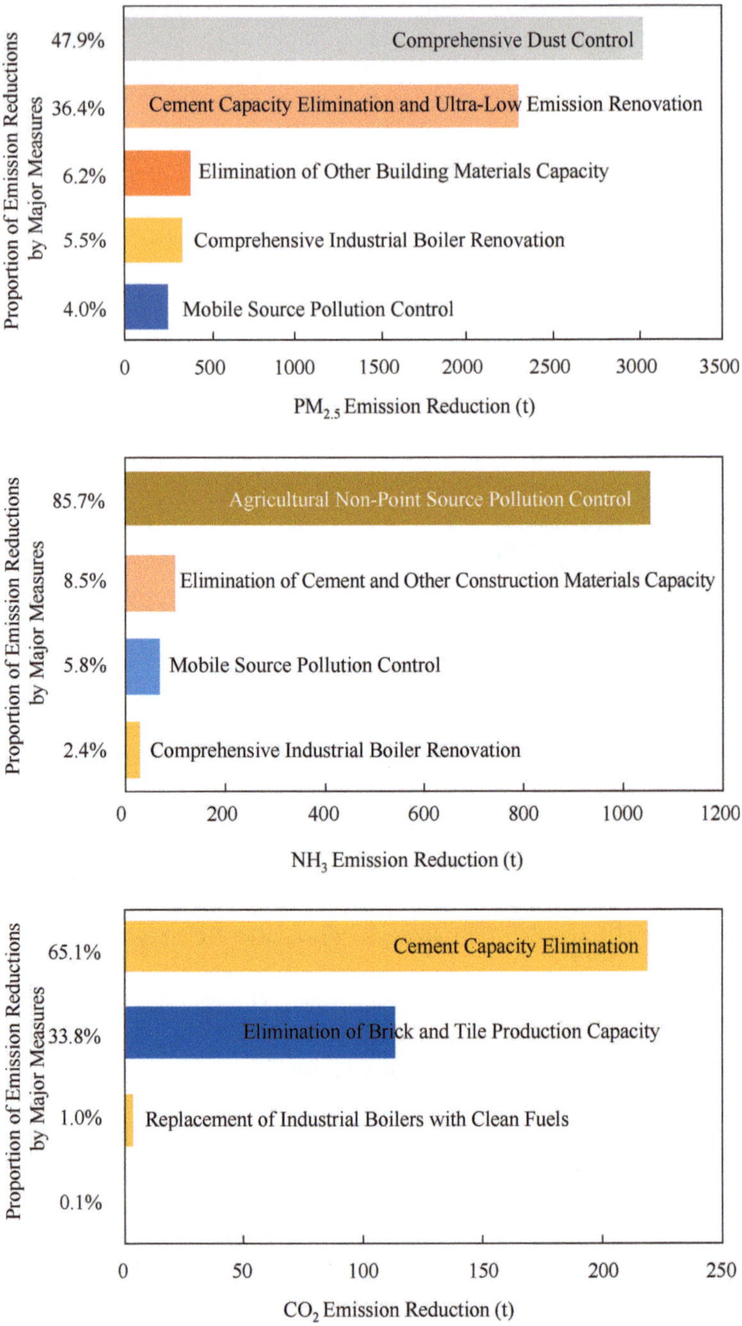

Fig. 5.3 (continued)

5.6 Analysis of Air Quality Compliance

Utilizing the iterative calculation method of the atmospheric environmental capacity within the WRF-CMAQ air quality model, simulations are conducted for sulfate, nitrate, ammonium, and the average proportion of primary $PM_{2.5}$ to $PM_{2.5}$ in Huzhou City. With the objective of meeting the annual average concentration standard for $PM_{2.5}$, emission reduction plans for SO_2, NO_x, NH_3, primary $PM_{2.5}$, and VOCs are devised. Through the implementation of spatially differentiated emission reduction strategies, new emission inventories are iteratively developed, and numerical simulations are conducted iteratively. The results obtained from iterative simulations reveal that by reducing local emissions of SO_2, NO_x, NH_3, primary $PM_{2.5}$, and VOCs by 26.7%, 28.6%, 10.0%, 28.9%, and 20.0%, respectively, relative to the baseline emissions in 2017, the simulated $PM_{2.5}$ concentration falls below the compliance limit of 35.0 $\mu g/m^3$. Consequently, based on the assessment of emission reduction measures, it is anticipated that atmospheric pollutant emissions in Huzhou City will align with the atmospheric environmental capacity under compliance conditions by 2020.

Based on the provided local pollutant reduction ratios, an emission inventory for Huzhou City in 2020 is established. Considering both local and neighboring regional emission reduction impacts, an analysis of the annual average simulated concentrations of major atmospheric pollutants in Huzhou City for the year 2020 is conducted. According to simulation outcomes, with the implementation of pollution control measures, it is anticipated that by 2020, the $PM_{2.5}$ concentration will decrease by 23.7% compared to 2017 levels, reaching around 32.2 $\mu g/m^3$. Specifically, considering only the impact of local emission reductions, the $PM_{2.5}$ concentration is expected to decline by 8.3 $\mu g/m^3$, reaching approximately 33.9 $\mu g/m^3$. Additionally, the reduction in $PM_{2.5}$ concentration due to diminished surrounding pollution transport is estimated to be 1.7 $\mu g/m^3$. For other pollutants, the 90th percentile of the maximum daily 8-hour O_3 concentration is expected to decrease by 3.1% relative to 2017 levels, with an estimated concentration of 180.8 $\mu g/m^3$ in 2020. The annual average simulated concentrations of PM_{10}, NO_2, and SO_2 are projected to decrease by 28.6%, 17.9%, and 11.8%, respectively, compared to 2017 levels, with estimated concentrations of 45.6, 31.4, and 13.0 $\mu g/m^3$ in 2020, respectively. Additionally, the 95th percentile of daily CO is anticipated to undergo a decline of 4.1% relative to 2017 levels, with an estimated concentration of 1.2 mg/m^3 in 2020. Therefore, by the year 2020, the major atmospheric pollutants, including $PM_{2.5}$, PM_{10}, SO_2, NO_2, and CO, in Huzhou are expected to meet the Grade II national standards for air quality, and O_3 pollution is not expected to show significant deterioration.

By the year 2025, through the comprehensive and profound optimization of energy, industrial, and transportation structures, along with the widespread elimination of old vehicles, the promotion and application of new energy vehicles, and the comprehensive substitution of low-VOC raw materials, concerted efforts will be directed towards facilitating regional joint prevention and control of air pollution. This involves innovating environmental management policies and measures, as well as enhancing the proactive pollution control initiatives of enterprises. With the implementation of these initiatives, it is anticipated that the O_3 concentration in Huzhou City will meet the Grade II national standards for ambient air quality.

5.7 Analysis of Predictions for Peak Carbon Emissions

The long-term carbon emission scenarios, as simulated and predicted by the LEAP model for Huzhou City, are visually depicted in Fig. 5.4. These scenarios incorporate the indirect carbon emissions associated with externally purchased electricity based on the emission inventory. In the BAU scenario, the transformation of energy structures and the adoption of energy-saving technologies in various sectors proceed at a relatively slow pace. However, with the overall restructuring of the industrial sector, a reduction in the proportion of energy-intensive industries, and an increase in external electricity purchases, the growth rate of carbon emissions is expected to decelerate. Ultimately, carbon emissions are anticipated to peak around 2030. In the AQP scenario, aiming for compliance by 2020 involves the closure of several cement and brick enterprises in Huzhou City between 2017 and 2020. The cement industry, characterized by high energy consumption and carbon emissions from fuel combustion and production processes, is targeted for significant carbon emission reduction benefits. This measure is expected to result in the peak of carbon emissions in Huzhou City as early as 2017, with carbon emissions reaching 95.6 million tons. Beyond 2020, carbon emissions in Huzhou City are anticipated to remain relatively stable, showing no significant increase, and a gradual decline is projected starting from 2028. Compared to the carbon emissions in the BAU scenario, carbon emissions in Huzhou City are projected to decrease by 3.46, 3.82, and 4.55 million tons in the years 2020, 2025, and 2030, respectively, under the AQP scenario.

The sector-specific carbon emissions for Huzhou City in the AQP scenario are elucidated in Fig. 5.5.

By the year 2020, the industrial sector in Huzhou City is anticipated to achieve a carbon emission reduction of 3.34 million tons compared to the baseline scenario.

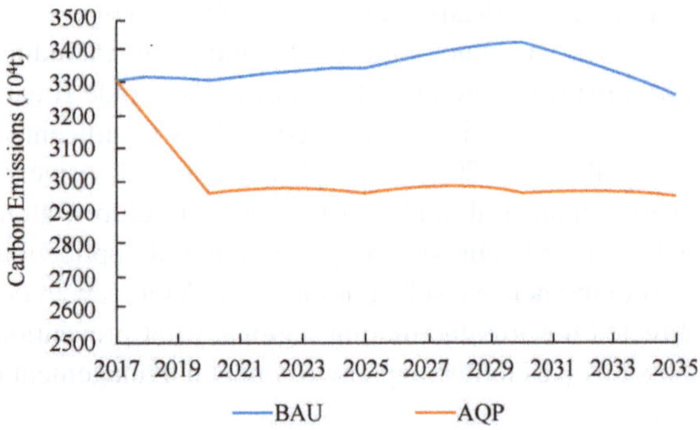

Fig. 5.4 Long-term carbon emission scenarios for Huzhou City simulated using the LEAP model

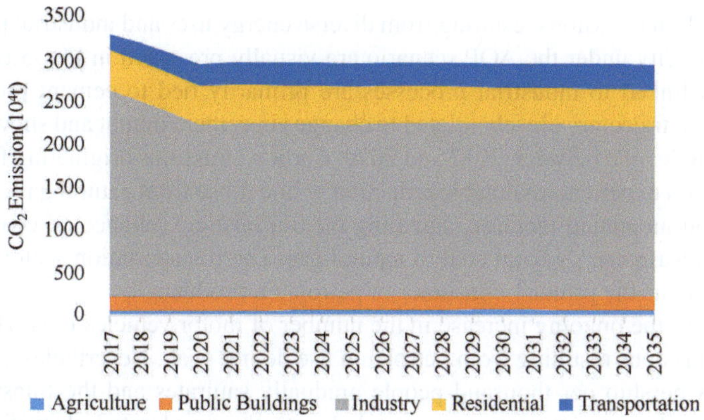

Fig. 5.5 Carbon emissions by sector in Huzhou City under the AQP scenario

Notably, the key contributing industries to the reduction of carbon emissions in industrial end terminals include non-metallic mineral manufacturing and construction. With the gradual increase in the proportion of high-tech industries in the future, encompassing biomedicine, new energy and energy conservation, new materials, opto-mechatronics, electronic information, and resource and environmental sectors, coupled with the upgrading of industrial energy-saving technologies, carbon emissions from the industrial sector in Huzhou City are expected to maintain a steady decline after experiencing a noticeable drop from 2017 to 2020.

Within the electricity sector, the AQP scenario envisions a sustained rise in the proportion of externally purchased electricity. Simultaneously, there will be a notable increase in the share of electricity generated from clean sources, including solar, hydro, wind, and other clean energy sources. As a result, carbon emissions from the electricity sector are anticipated to exhibit a continuous decrease, commencing from the year 2017.

Air quality policies exhibit a relatively constrained influence on emissions reduction within the public building sector, resulting in a reduction of carbon emissions by 76,000 tons by the year 2020 in comparison to the baseline scenario. Nevertheless, both energy demand and carbon emissions do not achieve their zenith. Owing to Huzhou's geographical placement in a region characterized by cold winters and hot summers, the long-term heating demand remains marginal. The projected upsurge in carbon emissions within the civilian sector is predominantly ascribed to the escalating demand for electricity for civil usage.

In the transportation sector, ongoing efforts will be dedicated to advancing the electrification process and restructuring transportation, particularly focusing on freight and public vehicles. However, the impact on emission reduction is moderate, resulting in a carbon reduction of 21,000 tons. The anticipated increase in the number of private cars is expected to contribute to a continuous elevation in overall energy demand and carbon emissions within the transportation sector.

The carbon emissions resulting from diverse energy uses and industrial processes in Huzhou City under the AQP scenario are visually presented in Fig. 5.6. Carbon emissions linked to industrial processes are primarily tied to cement production, displaying a trajectory closely related to changes in cement output and showcasing a substantial decline between 2017 and 2020. Carbon emissions originating from coal consumption experience a notable reduction, while those from natural gas consumption exhibit an annual increase, signaling the diminishing reliance on coal and the shift of end-use energy from coal to natural gas. The transportation sector emerges prominently as the primary consumer of petroleum products.

Owing to the ongoing increase in the number of motor vehicles in Huzhou City, carbon emissions resulting from petroleum use do not peak. Nevertheless, as motor vehicle ownership per thousand people gradually saturates and the transportation structure undergoes improvement, coupled with the accelerated electrification of freight and public vehicles, the growth rate of carbon emissions from petroleum use gradually decelerates. In the AQP scenario, the purchased electricity in Huzhou City is projected to surge from 2 billion kW·h in 2017 to approximately 5.6 billion kW·h by 2030, with the share of indirect carbon emissions from purchased electricity witnessing a gradual increase.

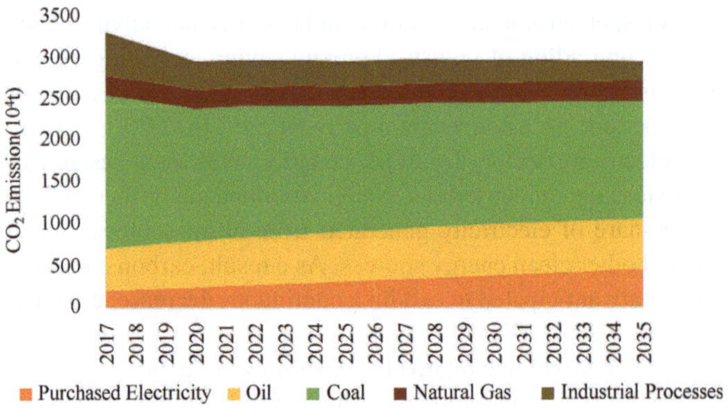

Fig. 5.6 Carbon emissions from different energy uses and industrial processes in Huzhou City under the AQP scenario

Chapter 6
Conclusions and Discussion

Abstract This book introduces a methodology termed "economy-energy-emission" coupled dynamic response analysis, which forms the basis for analysis in three selected cities—Zhengzhou, Shijiazhuang, and Huzhou. These cities are chosen as representative cases to examine structural adjustments and governance pathways across comprehensive service-oriented cities, industrial cities, and high-tech cities. Additionally, the study analyzes the co-benefits associated with various measures. The analysis reveals that while end-of-pipe control measures are pivotal in ensuring air quality compliance, they often lack coordinated effects on carbon emission reduction and may even yield adverse consequences. To achieve the objectives of "dual attainments", alongside end-of-pipe control measures, profound adjustments in energy, industrial, and transportation structures are deemed essential. Industrial sectors and electricity/heat production sectors, driven by air quality objectives, are identified as having the greatest potential for co-benefits in carbon emissions reduction. Policies such as electricity restructuring and the elimination of outdated production capacity are prioritized measures for each city due to their significant co-benefits for carbon emissions reduction. Furthermore, initiatives encompassing total control, industrial energy-saving renovations, elimination of coal-fired boilers, adoption of clean fuels in the civil sector, and restructuring in the transportation sector are highlighted as coordinated measures capable of simultaneously enhancing air quality and reducing CO_2 emissions.

The profound adjustment of the four major structures stands out as a pivotal driver for fostering green and low-carbon development. Initiatives encompassing total energy consumption control, electricity restructuring, industrial energy conservation and emission reduction, elimination of outdated production capacity, clean fuel adoption in the civil sector, and augmentation of green carbon sinks demonstrate significant co-benefits. A comprehensive examination of factors influencing the timing of "dual attainments" has been conducted, revealing discernible patterns with an expanding scope of studied cities. Recommendations for formulating "dual attainments" policies are posited, drawing upon emission inventories, urban characteristics, and the feasibility of proposed measures. However, it is acknowledged

Y. Bo et al., *Pathways for Reaching Air Quality Standard and Carbon Peak in Selected Cities in China*, Urban Sustainability, https://doi.org/10.1007/978-981-96-8315-4_6

that the research process entails inherent uncertainty, necessitating ongoing exploration and localized advancement of technologies such as emission inventories and air quality simulation to bolster the precision of research outcomes.

6.1 Summary of Methods

This investigation introduces a method for the coupled dynamic response analysis of the "economy-energy-emission" nexus and delineates a comprehensive framework for the coordinated examination of urban air quality compliance and the mitigation of peak carbon emissions. Notably, three cities, namely Zhengzhou City, Shijiazhuang City, and Huzhou City, have been chosen as representative cases for this study. The coordinated analysis framework commences with the compilation of emission inventories pertaining to urban air pollutants and greenhouse gases. It integrates various models, including the WRF-CMAQ air quality model and the LEAP model, to facilitate the computation of energy consumption and carbon emissions, accounting for policy and technological interventions. Furthermore, this framework enables an in-depth exploration of urban air quality conditions and variations in carbon emissions for the specified target year. Subsequently, it formulates customized strategies aimed at achieving the dual objectives of urban environmental compliance and carbon emissions reduction.

Presently in China, regulatory frameworks governing urban air quality control exhibit greater specificity and comprehensiveness compared to those addressing urban carbon emissions. Consequently, commencing with an overview of the selected case cities, the prevailing status of ambient air quality, and the characteristics of atmospheric pollutant emissions, this study delineates a timeline for achieving urban air quality compliance. Special emphasis is placed on meeting the $PM_{2.5}$ concentration standards essential for enhancing air quality. Key challenges in attaining air quality compliance are discerned through a meticulous examination of urban emission inventories and simulation outcomes. Utilizing an iterative calculation method, the study establishes maximum allowable emissions for each pollutant and devises emission reduction strategies in alignment with the outcomes of the emission source analysis. Subsequently, prospective energy technology scenarios and corresponding carbon emission levels are ascertained. Upon confirming the feasibility of the set objectives, pathways towards achieving urban air quality compliance are discerned. Simultaneously, an analysis of carbon emissions within this context reveals the scenario of peak carbon emissions under the guidance of air quality objectives. Furthermore, the carbon reduction benefits of each sector are subjected to a detailed analysis. Building upon the aforementioned analyses and findings from on-site research, this study proffers multifaceted policy recommendations for realizing air quality compliance and achieving peak carbon emissions in the case cities. These recommendations encompass the restructuring of industrial, energy, and transportation sectors, in addition to the incorporation of pollution control technologies.

6.2 Conclusions and Implications

6.2.1 "Dual Attainment" Strategies for Categorized Cities

6.2.1.1 Comprehensive Service-Oriented City

Zhengzhou is classified as a comprehensive service-oriented city. While exhibiting a relatively low carbon emission per unit of GDP, the city demonstrates robust economic development and sustains a significant population, thereby contributing to a comparatively higher level of absolute carbon emissions. In light of these circumstances, it becomes judicious to consider the formulation of an overarching total carbon emission control target. Noteworthy counterparts within this classification encompass prominent cities such as Beijing, Tianjin, Shanghai, Chongqing, Guangzhou, Shenzhen, Nanjing, Hangzhou, and Taiyuan, many of which function as either municipalities directly under the Central Government or provincial capitals.

Within comprehensive service-oriented cities, the industrial sector assumes a predominant role in both energy consumption and emissions. Coordinated emission reduction endeavors are primarily concentrated on energy production industries that supply secondary energy sources, including electricity and heating, as well as specific heavy industries. Strategic measures encompass the phase-out of obsolete infrastructure, the implementation of technological advancements, and the adoption of clean energy alternatives. In terms of industrial structure, emphasis should persistently be placed on fortifying the pivotal role of industrial scientific and technological innovation. The transition towards industrial intelligence and digitization warrants expeditious facilitation, with dedicated efforts directed towards cultivating advanced manufacturing clusters in key sectors such as biomedicine, next-generation information technology, artificial intelligence, new materials, new energy, and fields associated with energy conservation and environmental protection. Simultaneously, there is a requisite to exploit the substantial potential for energy conservation and efficiency enhancement. Targeted emission reduction initiatives should be directed towards critical areas, including transportation and construction.

Within comprehensive service-oriented cities, the civil sector stands out for its extensively electrified urban energy terminals, contributing to the reduced per capita pollutant emissions. The primary focus in this domain should be directed towards fostering a culture of energy-saving lifestyles among residents and overseeing the judicious management of overall energy consumption. Additionally, it remains imperative to consistently advocate for the replacement of civilian bulk coal in suburban and rural areas with cleaner energy alternatives, employing measures such as subsidies to facilitate this transition.

In the transportation sector of comprehensive service-oriented cities characterized by high population density and situated within air quality control areas, the implementation of measures such as "vehicle purchase restrictions" serves to regulate the overall quantity of motor vehicles. The promotion of public transportation usage and the enforcement of daily "vehicle use restrictions" represent additional strategies to mitigate traffic congestion effectively. Furthermore, expediting the adoption of subsequent phases of motor vehicle emission standards and advanced fuel economy standards is advisable, facilitating the gradual phase-out of antiquated, highly polluting, and energy-intensive vehicles. Concomitant subsidy policies can be introduced to incentivize the widespread adoption of new energy vehicles within both public and civilian domains. Simultaneously, efforts should be directed towards augmenting infrastructure development, particularly the expansion of charging stations.

6.2.1.2 Industrial Production-Oriented City

Cities oriented towards industrial production can be further classified into two distinct categories: heavy industrial cities and light industrial cities.

1. Heavy industrial cities.

   Shijiazhuang City is categorized as an industrial city predominantly focused on heavy industry, with its pivotal sector centered on the production of primary materials that constitute the foundational materials for various segments of the national economy. The city accommodates a significant concentration of high-energy-consuming enterprises, substantial coal consumption, and comparatively elevated carbon emissions per unit of GDP. This scenario contributes substantially to formidable challenges in air quality control. Analogous cities with comparable industrial characteristics encompass Tangshan, Handan, Baotou, Chifeng, Ma'anshan, Luoyang, Changchun, and Hanzhong.

   In the future, heavy industrial cities are urged to align closely with the imperatives of constructing a resource-efficient and environmentally friendly society. This necessitates expeditious advancements in technologies and equipment dedicated to energy conservation, environmental protection, and resource recycling. Within the framework of supply-side structural reform, these cities should proactively cultivate novel driving forces and capitalize on emerging opportunities to facilitate transformative development.

   Within the industrial sector, heavy industrial cities ought to prioritize addressing the infrastructure of key industries, including cement, steel, chemicals, and non-ferrous metals, characterized by elevated levels of energy consumption and pollution. Strategic interventions such as technological upgrades, energy-saving transformations, capacity reduction, and the phase-out of obsolete capacities can collectively contribute to diminishing the energy consumption intensity of pivotal heavy industrial infrastructure. Renovation initiatives targeting industrial boilers, the implementation of energy-saving measures in motor systems, and the harnessing of industrial waste heat represent effective measures to enhance the energy efficiency of these foundational industries. Moreover, heavy industrial cities should focus on improving the processing and conversion efficiency within the electricity and heat production sectors. This involves the phased elimination of outdated small coal-fired power units, incentivizing industrial enterprises to transition from coal to gas in self-owned power plants, adopting combined heat and power systems, and establishing distributed energy stations.

In the civilian sector, heavy industrial cities should actively promote the replacement of civilian bulk coal with cleaner energy sources. Concurrently, initiatives aimed at residential energy-saving renovations, the improvement of residential insulation performance, and the promotion of modern, clean, and efficient coal-burning stoves are essential measures.

Within the transportation sector, heavy industrial cities should implement more stringent vehicle emissions standards, improve fuel efficiency, and actively endorse the adoption of clean energy vehicles, particularly in public transportation and corporate transport fleets. Given the substantial involvement of heavy materials transportation in sectors such as steel, construction materials, and chemicals, there is a critical need to emphasize and facilitate the transition from road-based transportation to railways and waterways. Concurrently, promoting multimodal transportation is imperative, as it optimizes the transportation of bulk goods.

2. Light industrial cities

Light industrial cities specialize in the provision of consumer goods and the production of handcrafted products, representing a traditional and advantageous industry for exports in China. In contrast to heavy industry and energy production sectors, light industrial cities generally demonstrate lower levels of energy consumption. Among the myriad light industrial cities in China are notable examples such as Weifang, Putian, Xinxiang, Chaozhou, Luzhou, Bayannur, and Yanbian Korean Autonomous Prefecture.

In the industrial sector, light industrial cities should endeavor to stabilize the development of traditional products while actively promoting technological upgrades within the light industry. Simultaneously, there is a pressing need to expedite the development of high-tech industries in the future. Coordinated emission reduction initiatives in these cities should prioritize key sectors, including energy production (primarily electricity and heat supply), heavy industry, and pivotal segments within the light industry. To this end, it is advisable to encourage industrial enterprises to transition from coal to gas in self-owned power plants, embrace combined heat and power generation, and establish distributed energy stations as strategic measures. Given the relatively low entry barriers in the light industry and an oversupply of production capacity for numerous products, there exists a risk of unregulated "scattered, non-compliant, and polluting" enterprises. Therefore, promoting the clustering of light industry becomes imperative to enhance production efficiency, reduce costs, and minimize energy consumption.

In the civilian sector, light industrial cities should actively advocate for the substitution of civilian bulk coal with cleaner energy sources. Moreover, initiatives focusing on residential energy-saving renovations, improvements in residential insulation performance, and the promotion of modern, clean, and efficient coal-burning stoves are integral components of sustainable urban living practices.

Within the transportation sector, light industrial cities are urged to enforce more stringent vehicle emission standards, improve fuel efficiency, and actively endorse the adoption of clean energy vehicles, particularly within public fleets.

6.3 High-Tech Cities

Huzhou City is classified as a high-tech city, characterized by a substantial presence of the secondary and tertiary sectors, particularly featuring a notable share of high-tech industrial added value within its industrial sector. In terms of carbon emissions, both the total volume and carbon intensity in this city are relatively low. High-tech industries in Huzhou, marked by their technology-intensive and capital-intensive nature, exhibit significantly lower levels of energy consumption intensity compared to other industries. Similar high-tech cities, such as Wuxi, Changzhou, Suzhou, Yangzhou, Zhongshan, Huizhou, and Foshan, are primarily situated in economically developed and highly open regions like the Pearl River Delta and the Yangtze River Delta.

In the realm of energy, high-tech cities should actively promote the transition from fossil fuels to non-fossil fuels across various sectors, including transportation, energy, industry, and construction. A crucial aspect involves significantly increasing the electrification level of energy consumption terminals and establishing an innovative energy system with electricity as the central focus, complemented by diverse forms of energy sources. The realization of a "low-carbon transformation" within the electricity industry stands as a paramount objective. Furthermore, high-tech industries should strive to elevate the proportion of renewable energy in their energy mix, intensify the integration of green electricity, implement electricity substitution across diverse sectors, and augment the overall share of electricity in terminal energy consumption.

In the industrial sector, high-tech cities should continue to increase investments in research and development, guiding enterprises towards transformative upgrades that embrace green, intelligent, and digital technologies. A concerted effort to promote technological progress and harness the cluster effects within the industry is essential. Additionally, there is a need to further elevate the proportion of new technology industries and advocate for the development of low-carbon sectors, including energy-efficient and environmentally friendly technologies, biopharmaceuticals, new energy, and new materials. These strategic initiatives play a pivotal role in improving emission source efficiency and advancing coordinated efforts for emission reductions.

In the domain of construction, high-tech cities should actively champion green building practices. This involves the incorporation of energy-efficient materials in construction projects, the installation of environmentally friendly electrical appliances, and the implementation of energy-saving retrofits for older buildings.

Regarding transportation, high-tech cities should prioritize green transportation as a developmental direction. This entails the enforcement of stringent emission standards for motor vehicles, continuous improvements in fuel efficiency, and a sustained increase in the adoption of new energy vehicles. Additionally, initiatives should be directed towards guiding the phased elimination and scrapping of old motor vehicles in an orderly manner, while reinforcing emission control measures for ships and non-road mobile machinery.

6.3.1 Common Experiences of Urban "Dual Attainments"

The "dual attainments" strategies implemented in the case cities—Zhengzhou, Shijiazhuang, and Huzhou—have been summarized in Table 6.1, highlighting specific measures that contribute significantly to the reduction of carbon emissions corresponding to each pollutant. Key strategies encompass end-of-pipe control measures and the elimination of outdated production capacity in industrial enterprises, playing a pivotal role in reducing emissions of SO_2, NO_x, and $PM_{2.5}$. Notably, measures with a pronounced impact on CO_2 emissions reduction are primarily associated with the restructuring of the power sector and the elimination of outdated production capacity, resulting in a substantial decrease in the use of fossil fuels and emissions from industrial processes. Given the diverse sources of different pollutants, optimal emission reduction measures vary accordingly. Specifically, for VOCs, the most effective reduction measure involves targeted VOCs control, specifically addressing the reduction of VOCs emissions.

The measures falling under the end-of-pipe control category, such as industrial upgrading and transformation, serve as essential safeguards for achieving air quality compliance. However, these measures often do not yield significant co-benefits in carbon emissions reduction. To successfully realize the objectives of "dual attainments", it is imperative to complement end-of-pipe control measures with a profound restructuring of energy, industrial, and transportation systems. This holistic approach presents an opportunity to cultivate new dynamics for urban economic growth and

Table 6.1 Specific measures with the significant carbon emission reduction contributions for each pollutant

Pollutant	Major Measures Contributing to Emission Reductions
So_2	Industrial upgrading and renovation, elimination of outdated capacities in key industries, fuel cleaning and substitution
NO_x	Mobile source pollution control, industrial upgrading and renovation, elimination of outdated capacities in key industries
VOCs	Volatile organic compounds control
$Pm_{2.5}$	Comprehensive dust control, resolving excess production capacity, industrial upgrading and renovation
NH_3	Agricultural non-point source pollution control
CO_2	Power structure adjustment, elimination of outdated capacities in key industries

drive the transition towards a green and low-carbon economy. This transition stands as a crucial driving force for the sustainable development of cities in the future. Significantly, the industrial sector, in tandem with the electricity and heat production sectors, holds the potential for the most substantial co-benefits in carbon emissions reduction. Policies focusing on restructuring in the power sector and the elimination of outdated production capacities deliver notable co-benefits in carbon emissions reduction and should be prioritized by cities aiming for dual attainments.

The synthesis of coordinated policies, aimed at concurrently ameliorating air quality and effecting a reduction in CO_2 emissions, is delineated in Table 6.2. These strategies encompass total control measures, restructuring initiatives in the power sector, implementation of industrial energy-saving transformations, phased elimination of outdated production capacities, substitution of coal-fired boilers, promotion of clean fuel usage in the civil sector, restructuring endeavors in the transportation sector, and augmentation initiatives for green carbon sinks. It is imperative to acknowledge that the outcomes of emission reduction policies associated with the procurement of external electricity and the electrification of motor vehicles are intricately tied to the prevailing power structure. A preponderance of thermal power does not necessarily yield co-benefits, underscoring the pivotal role of clean electricity. For instance, the policy of procuring external electricity may effectively diminish local pollutant emissions in cities but could concurrently engender an escalation of pollutant emissions from power-producing cities. Furthermore, given the current conversion efficiency of thermal power being less than 50%, a power structure predominantly characterized by thermal power in power-producing cities may paradoxically result in an overall increase in carbon emissions.

The contribution of emission reduction from two-tier emission sources in the three case cities is elucidated in Fig. 6.1. In alignment with air quality objectives, the industrial sector and power heating sector emerge as the sectors exhibiting the most substantial co-benefits for carbon emissions reduction. Measures such as the restructuring in the industrial sector, reduction of high-energy-consuming product output, and advancements in industrial technology generate noteworthy co-benefits, constituting over 80% of the overall carbon emissions reduction. While prevailing air quality policies for the civilian and transportation sectors emphasize stricter emission standards and the promotion of the electrification process, resulting in discernible improvements in air quality, the co-benefits for carbon emissions reduction remain relatively limited. Considering the enhancements in people's living standards, population growth, and the prospective development of the logistics industry, the energy demand for the urban civilian and transportation sectors is anticipated to stand at elevated levels for an extended period. Consequently, this trajectory is poised to gradually augment their share of carbon emissions in the overall carbon emissions profile.

In the forthcoming future, the potential for carbon emissions reduction in urban industrial sectors, relying on the advancement of energy-saving technologies and the reduction of production capacities, is expected to gradually diminish. The emphasis of carbon reduction initiatives will pivot towards adjustments in industrial structure. Simultaneously, cities should heighten their efforts to regulate the overall energy

Table 6.2 Policies with carbon emission reduction co-benefits

Department	Field	Specific Policy	Synergistic Benefits?
Electricity	Total control	Strict control of new coal-fired unit installation	+
		Accelerated phasing out of coal-fired units below 300,000 kW	+
		Increase in renewable energy generation capacity to replace local coal-fired power production	+
		Increase the import of green electricity to reduce local coal-fired power generation	+
		Replacement of coal-fired units with gas-fired units	+
		Phasing out corporate-owned coal-fired units	+
	Energy efficiency	Optimize power plant operational dispatch, high-efficiency plants share the load of low-consumption plants	+
		Accelerate the research, development, and application of high-efficiency power generation technologies, reduce coal consumption per unit of electricity supplied	+
	End-control	Ultra-low emission retrofitting of power plants	-
		Intensified management of unorganized emissions at power plants	–
		Research, development, and application of carbon capture and utilization (CCU) technologies at power plants	/
Industry	Industrial structure adjustment	Strictly control the environmental entry of high-pollution, high-energy-consuming industries	+
		Eliminate outdated and inefficient capacities, reduce excess capacity	+
		Rectify "scattered, polluted, and disorderly" enterprises	+
		Develop energy-saving, environmental protection, new energy, and other green low-carbon industries	+

(continued)

Table 6.2 (continued)

Department	Field	Specific Policy	Synergistic Benefits?
	Energy structure optimization	Strictly control new coal-fired projects	+
		Eliminate small coal-fired boilers	+
		Substitute coal-fired boilers with electric power	+
		Substitute coal-fired boilers with natural gas	+
		Implement substitution of coal with electricity in industrial furnaces and kilns	+
		Implement substitution of coal with natural gas in industrial furnaces and kilns	+
		Construct "no-burn zones" for high-pollution fuels	+
		Improve energy efficiency in key industries	+
		Strictly control coal quality to ensure sulfur content meets standards	/
	Source control	Strengthen control of VOCs at the source of raw and auxiliary materials	/
	End control	Strengthen end control of boilers	-
		Implement upgrading and standard enhancement for industrial furnaces and kilns	–
		Implement upgrading and standard enhancement in key industrial sectors	–
		Strengthen the management of unorganized emissions from industrial enterprises	–
	Heavy pollution emergency	Seasonal pollution emission control. Implement emergency emission reduction projects for heavy pollution	+
Transportation	Total control	Control the number of vehicles owned	+
	Source control	Tighten emission standards for vehicles and non-road mobile machinery	/
		Enhance the quality upgrade and supervision of fuel	/
		Reduce the intensity of vehicle use	+
		Phase out diesel trucks meeting only emission standards III or lower	+
	End control	Promote in-depth management of high-emission vehicles, non-road mobile machinery, and ships	/
		Strengthen vapor recovery management and supervision	–

(continued)

Table 6.2 (continued)

Department	Field	Specific Policy	Synergistic Benefits?
	Improve efficiency	Improve fuel economy, reduce average fuel consumption	+
	Optimize energy structure	Promote the adoption of new energy vehicles, non-road mobile machinery, and ships	+
	Transport structure adjustment	Increase the proportion of rail and water freight transport	+
		Develop multimodal transportation	+
		Build a green urban logistics system	+
		Improve public transportation and increase the public transit ridership rate	+
		Develop and promote "low-emission driving zones"	+
Construction	Improve efficiency	Fully implement green building design standards	+
		Actively develop ultra-low energy buildings	+
		Implement energy-saving renovations in existing buildings	+
		Widely promote prefabricated buildings	+
Residential	Source control	Advocate for a low-carbon lifestyle. Implement subsidies for energy-saving products	+
	Improve efficiency	Promote new types of clean, high-efficiency, energy-saving stoves.	+
	Optimize energy structure	Implement "coal to electricity" for winter heating	+
		Implement "coal to gas" for winter heating	+
		Implement "coal to centralized heating" for winter heating	+
		Implement "coal to renewable energy" for winter heating	+
		Accelerate the replacement of coal with electricity in agricultural facilities and service industries	+
		Accelerate the replacement of coal with natural gas in agricultural facilities and service industries	+
		Improve coal quality, use low-ash, low-sulfur clean coal and briquettes	+
	End control	Encourage in-depth treatment of cooking fumes, install high-efficiency fume purification devices	–

(continued)

Table 6.2 (continued)

Department	Field	Specific Policy	Synergistic Benefits?
Land use	Land structure adjustment	Comprehensive dust control	/
		Control of ammonia emissions from agricultural sources	/
		Ban on straw burning	+
		Promotion of greening and carbon sequestration projects	+

"+" signifies policies yielding co-benefits for both the enhancement of air quality and the reduction of carbon emissions.
"–" indicates policies that provide benefits for air quality enhancement but concurrently have negative effects on carbon emissions reduction.
"/" represents policies affording benefits in either air quality enhancement or carbon emissions reduction.
* Additionally, it is noteworthy that the co-benefits associated with the promotion of electric vehicles are contingent upon the prevailing power structure. In instances where thermal power predominates within the power structure, there is a potential for heightened SO_2 emissions [1]

demand within civilian and transportation sectors, while concurrently fostering the low-carbon transformation of fuel structures. Neglecting these measures may pose a risk of a rebound in carbon emissions.

The computed CO_2 values exclusively signify direct emissions and do not encompass allocated values for secondary energy distribution in electricity and heat.

Within the purview of this study, it is anticipated that the peak carbon emissions for the three representative case cities will precede the achievement of air quality compliance ($PM_{2.5}$ concentration < 35 μg/m^3), as elucidated in Table 6.3. The industrial sector emerges as the primary contributor to peak carbon emissions in typical case cities, primarily due to relatively lenient carbon reduction policies in the civilian and transportation sectors. For instance, Huzhou City implemented the closure of cement and brick factories, Shijiazhuang City reduced steel production capacity, and Zhengzhou City curtailed cement and carbon production capacity. In the short to medium term, all three typical case cities promptly curtailed the utilization of fossil fuels, particularly coal, thereby expediting the achievement of peak carbon emissions.

Existing data also reveals that certain cities exhibit disparate characteristics. For example, in 2014, the $PM_{2.5}$ concentration in Shenzhen City already met the Grade II national standards for air quality, yet it is anticipated to achieve peak carbon emissions only by 2020 [2]. Urban air quality compliance establishes unequivocal requirements for diverse pollutant emissions, whereas peak carbon emissions impose requirements on the relative change (year-on-year decline) in carbon emissions. The timeline for accomplishing the "dual attainments" objectives is influenced by various factors, encompassing the current air quality status, economic development, and emission reduction measures implemented in the city. Owing to the limited number of case cities examined, systematic correlations between the timelines for urban peak carbon

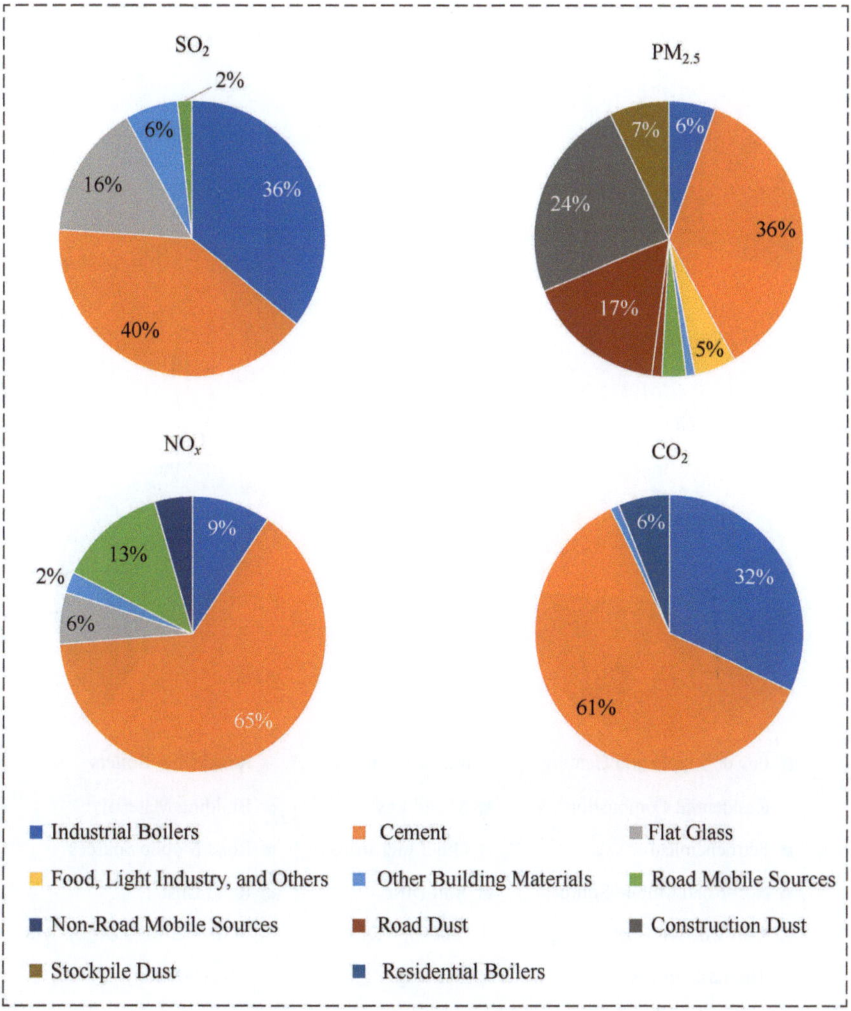

(a) Huzhou

Fig. 6.1 Emission reduction contributions from two-level emission sources in three case cities

emissions and air quality compliance cannot be conclusively drawn at present. This underscores the need for further research in the future.

Focused on the comprehensive formulation of policies to realize urban "dual attainments" objectives, this book advances the following recommendations.

3. Emission inventories serve as the foundational element for research directed at achieving the "dual attainments" objectives. To ensure uniformity in data statistical specification, cities are advised to integrate their monitoring and statistical

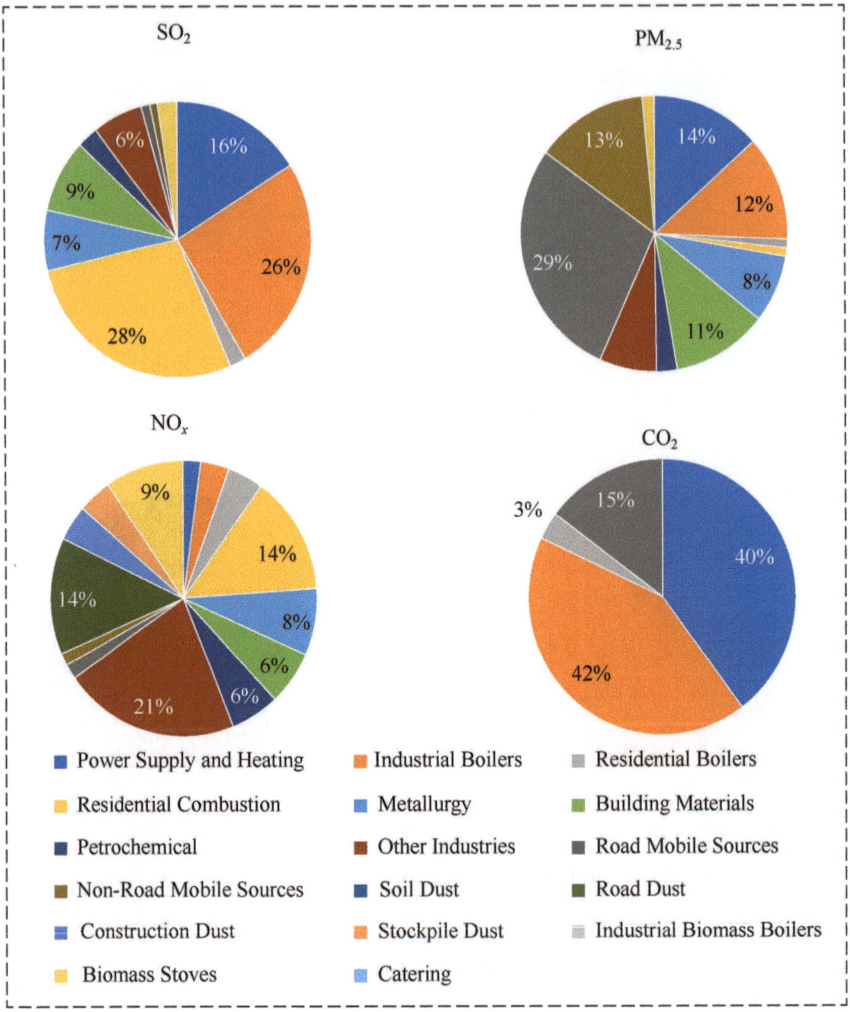

(b) Shijiazhuang

Fig. 6.1 (continued)

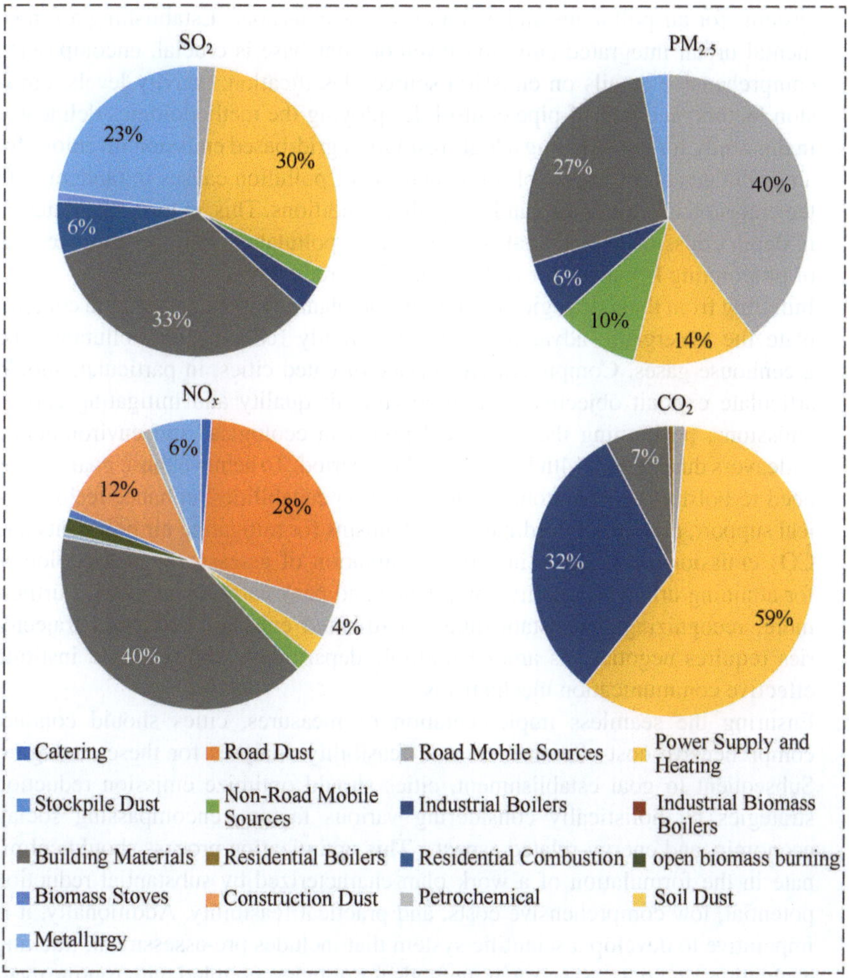

(c) Zhengzhou

Fig. 6.1 (continued)

Table 6.3 Timeline for "dual attainments" in case cities

City	Estimated Year of Carbon Emission Peaking	Year of Air Quality Compliance
Huzhou	2017	2020
Zhengzhou	2025	2028
Shijiazhuang	2020	2033

systems for air pollutants and greenhouse gas emissions. Establishing a fundamental urban integrated emission inventory database is crucial, encompassing comprehensive details on emission source classification, activity levels, emission factors, and end-of-pipe control. Employing the methodologies delineated in this study for constructing a high-resolution grid-based emission inventory for air pollutants and CO_2 enables the analysis of pollution causes in tandem with the emission inventory data and air quality conditions. This, in turn, facilitates an in-depth emission source analysis for various pollutants, with the ultimate goal of pinpointing key areas for carbon emissions reduction.

4. Initiating from their strategic positioning and planning, the cities should contemplate the synergistic advantages of concurrently reducing air pollutants and greenhouse gases. Comprehensive service-oriented cities, in particular, should articulate explicit objectives for improving air quality and mitigating carbon emissions, positioning them as focal points in ecological and environmental endeavors during the "14th Five-Year Plan" period. To achieve these goals, cities need to bolster their environmental protection capabilities, enhance technological support, prioritize coordinated mechanisms for mitigating air pollutants and CO_2 emissions, and delve into the formulation of assessment methodologies for attaining urban air quality compliance and peak carbon emissions. Furthermore, recognizing that establishing coordinated emission reduction trajectories requires negotiations among multiple departments, cities should institute effective communication mechanisms.
5. Ensuring the seamless implementation of measures, cities should conduct comprehensive cost-effectiveness and feasibility analyses for these strategies. Subsequent to goal establishment, cities should optimize emission reduction strategies by holistically considering various factors, encompassing social, economic, and energy-related aspects. This optimization process should culminate in the formulation of a work plan characterized by substantial reduction potential, low comprehensive costs, and practical feasibility. Additionally, it is imperative to develop a scientific system that includes pre-assessment, tracking evaluation, and post-assessment methods for policies, coupled with routine evaluations and strategy refinements. Drawing insights from the experiences of cities that have already undertaken such initiatives, challenges in urban "dual attainments" efforts are underscored by the elevated costs associated with end-of-pipe control and difficulties in adjusting industrial structure.
6. Strengthening pertinent systems and ensuring the efficacious implementation of policies are pivotal imperatives. Cities should fortify law enforcement and regulatory capabilities, propel advancements in grid-based environmental monitoring, and integrate carbon emission trading systems along with pollution source management systems, incorporating emission permits. This integration endeavors to curtail costs for both businesses and government administration. Furthermore, establishing coordinated assessment mechanisms and exploring integrated assessment methods within an overarching evaluation system is deemed essential.

7. Reinforcing regional coordinated emission reduction is imperative. The outcomes of atmospheric pollutant transport analysis underscore the substantial influence of surrounding areas on the transport of $PM_{2.5}$ concentrations in Zhengzhou, Shijiazhuang, and Huzhou, with annual average impact rates of 34.5%, 43.4%, and 32.2%, respectively. The attainment of air quality compliance in the target cities becomes challenging if regional coordinated emission reduction for air pollutants is not realized. Moreover, in the context of considering the procurement of external electricity, a reduction in the carbon emission coefficient of the regional power grid proves advantageous for carbon emissions reduction in the target cities. Therefore, regional coordinated emissions reduction is vital for cities to successfully achieve the "dual attainments" goals.

6.4 Uncertainty Discussion

The coordinated analysis framework for achieving "dual attainments", as delineated in this book, integrates diverse technical methodologies, including emission inventories, measure evaluations, emission forecasts, and numerical simulations. This comprehensive approach involves various foundational data types. It is essential to acknowledge that the entire analysis process inherently incorporates uncertainties. Consequently, two pivotal components, namely emission inventories and numerical simulations, are selected for explicit elaboration on the sources and attributions of uncertainty.

6.4.1 Uncertainties in Emission Inventories

In the compilation of emission inventories, the uncertainty of emission levels primarily emanates from the following aspects.

1. Data gaps: One of the most substantial sources of uncertainty in the compilation of emission inventories arises from data gaps. This is primarily manifested by the absence of reliable emission factor information and authentic activity level data.
2. Data representativeness: During the compilation of emission inventories, the emission factors or other parameters employed may not sufficiently represent the emission characteristics of the estimation object. Furthermore, the lack of representativeness in the sources of adopted data can contribute to uncertainty in the estimation of emission inventories.
3. Variability: During the compilation of emission inventories, average values derived from multiple emissions source tests are often used to depict the emission characteristics and levels of a specific category of emission sources. However, even within the same category of emission sources, variations in factors such as individual emission process levels, fuel types, and control levels can lead

to variability in emissions [3]. This variability significantly contributes to the uncertainty in estimating emission factors or emission inventories.

4. Non-standardized data sources: Challenges emerge during the compilation of emission inventories due to the absence of unified standards and regulations for data collection, inconsistencies in data usage, and the unreasonable selection of substitute data. The resulting quality issues in data sources significantly contribute to the uncertainty in emission inventories.

6.4.2 Uncertainties in Numerical Simulations

This book utilizes numerical simulation methods with an air quality model to perform concentration simulations of atmospheric pollutants, including $PM_{2.5}$, NO_2, SO_2, and O_3, in the baseline year. Subsequently, it assesses regional transport impacts, atmospheric environmental capacity, and the realizability of emission reduction strategies in the target planning year. Throughout the numerical simulation process, uncertainties in simulating atmospheric pollutant concentrations may arise from the local gridded emission inventory in the study area, the air quality model, the meteorological model, and the transport impact of emissions from surrounding areas. A detailed analysis of these factors is presented below.

1. Uncertainty arising from emission inventories: In the compilation of emission inventories, uncertainties in the numerical simulation results may stem from the omission of emission sources, as well as the absence or inadequate representativeness of key data such as emission factors and activity levels. Additionally, processes like spatio-temporal allocation and chemical species distribution contribute to uncertainties. For instance, the formation mechanism of $PM_{2.5}$ is highly intricate, and the emissions of various precursor components are challenging to estimate, significantly impacting the accurate simulation of $PM_{2.5}$. The generation of O_3 is influenced by NO_x or VOCs, and errors in the NO_x or VOCs emissions within the emission inventory will propagate into the simulation process of O_3 concentrations.
2. Uncertainty in the simulation of the air quality model: This book employs the WRF-CMAQ air quality model to simulate the air quality in Zhengzhou City, Shijiazhuang City, and Huzhou City. Among other factors, the complexity of the air quality model and numerous input parameters introduce uncertainties into the simulation of atmospheric physical and chemical processes. In other words, the results of the air quality model simulations inevitably contain a certain degree of error.
3. Uncertainty due to differences in meteorological conditions across different years: Certain differences exist in meteorological conditions between the target year (2028 for Zhengzhou, 2033 for Shijiazhuang, and 2020 for Huzhou) and the baseline year (2017). In predicting air quality for the target year, this book utilized meteorological conditions from the baseline year, introducing some error into the simulation results.

4. The transport impact of emissions from surrounding areas: Emissions from neighboring areas exert a certain transport impact on atmospheric pollutants in the study region. In the target year, there may be discrepancies in the emission structure and magnitude of atmospheric pollutants in the vicinity of the study area compared to the study assumptions, introducing uncertainty into the simulation results.

References

1. China Sustainable Transportation Center. Technical Guidelines for Coordinated Control of Air Pollutants and Greenhouse Gases in Urban Transportation Version 1.0 (in Chinese) [R]. Beijing: Energy Foundation, 2019. https://www.efchina.org/Reports-zh/report-lccp-20191212-zh.
2. Tang, J., et al. Achieving Triple Goals of Carbon Emission Peaking, Air Quality Standard Attainment, and Economic Prosperity at the City Level: the Shenzhen Case (in Chinese) [R]. Beijing: Energy Foundation, 2019. https://www.efchina.org/Reports-zh/report-lccp-20191015-zh.
3. Zheng, B., Huo, H., Zhang, Q., et al. High-resolution Mapping of Vehicle Emissions in China in 2008[J]. Atmospheric Chemistry and Physics, 2014, 14: 9787-9805.

4. The transport impact of emissions from surrounding areas: Emissions from neighb[illegible] [illegible] transport [illegible] on atmospheric pollutants in the study [illegible]

References

[illegible]

MIX
Papier aus verantwortungsvollen Quellen
Paper from responsible sources
FSC® C105338
FSC
www.fsc.org

If you have any concerns about our products,
you can contact us on
ProductSafety@springernature.com

In case Publisher is established outside the EU,
the EU authorized representative is:
Springer Nature Customer Service Center GmbH
Europaplatz 3, 69115 Heidelberg, Germany

Printed by Libri Plureos GmbH
in Hamburg, Germany